Lecture Notes in Computer Science 12102

More information about this series at http://www.springer.com/series/7407

Luís Paquete · Christine Zarges (Eds.)

Evolutionary Computation in Combinatorial Optimization

20th European Conference, EvoCOP 2020
Held as Part of EvoStar 2020
Seville, Spain, April 15–17, 2020
Proceedings

 Springer

Editors
Luís Paquete (iD)
University of Coimbra
Coimbra, Portugal

Christine Zarges (iD)
Aberystwyth University
Aberystwyth, UK

ISSN 0302-9743   ISSN 1611-3349   (electronic)
Lecture Notes in Computer Science
ISBN 978-3-030-43679-7   ISBN 978-3-030-43680-3   (eBook)
https://doi.org/10.1007/978-3-030-43680-3

LNCS Sublibrary: SL1 – Theoretical Computer Science and General Issues

This Springer imprint is published by the registered company Springer Nature Switzerland AG
The registered company address is: Gewerbestrasse 11, 6330 Cham, Switzerland

Preface

Combinatorial optimization is concerned with finding an optimal solution from a huge finite set of candidate solutions. Problems of this nature are omnipresent in many of today's societal, industrial, and scientific challenges and include important areas such as scheduling, timetabling, network design, transportation and distribution, vehicle routing, stringology, graphs, satisfiability, energy optimization, cutting, packing, planning, and search-based software engineering. However, classical optimization techniques often cannot cope with the complexities and constraints of these problems. Heuristic methods such as evolutionary algorithms and other nature-inspired approaches as well as advanced local search techniques constitute very powerful and successful approaches that are able to produce high-quality solutions in reasonable time. They are typically very easy to implement and apply and come with the advantage of being anytime algorithms, i. e., they can be stopped at any time during the optimization process and will return the best solution seen so far. This clearly distinguishes them from classical algorithms that usually do not return any solution if stopped prematurely. This way, anytime algorithms provide a very flexible and intuitive approach to problem-solving that can easily be configured to user needs such as time constraints. Recent theoretical and experimental advances in this area are the main topics of these proceedings.

This volume contains the proceedings of the 20th European Conference on Evolutionary Computation in Combinatorial Optimisation (EvoCOP 2020). The conference was held in Seville, Spain, from April 15–17, 2020. The EvoCOP conference series started in 2001, with the first workshop specifically devoted to evolutionary computation in combinatorial optimization. It became an annual conference in 2004. EvoCOP 2020 was organized together with EuroGP (the 23rd European Conference on Genetic Programming), EvoMUSART (the 9th International Conference on Computational Intelligence in Music, Sound, Art and Design), and EvoApplications (the 23rd European Conference on the Applications of Evolutionary Computation, formerly known as EvoWorkshops), in a joint event collectively known as EvoStar 2020. Previous EvoCOP proceedings were published by Springer in the *Lecture Notes in Computer Science* series (LNCS volumes 2037, 2279, 2611, 3004, 3448, 3906, 4446, 4972, 5482, 6022, 6622, 7245, 7832, 8600, 9026, 9595, 10197, 10782, and 11452). The table on the next page reports the statistics for each of the previous conferences.

This year, 14 out of 37 papers were accepted after a rigorous double-blind review process, resulting in a 38% acceptance rate. We would like to acknowledge the quality and timeliness of our Program Committee members' work. Decisions considered both the reviewers' report and the evaluation of the program chairs. The 14 accepted papers cover a wide spectrum of topics, ranging from the foundations of evolutionary computation algorithms and other search heuristics, to their accurate design and application to combinatorial optimization problems. Fundamental and methodological aspects deal with runtime analysis, the structural properties of fitness landscapes, the study of metaheuristics core components, the clever design of their search principles, and their careful selection and configuration. Applications cover problem domains such as scheduling, routing, partitioning, and general graph problems. We believe that the range of topics covered in this volume of EvoCOP proceedings reflects the current state of research in the fields of evolutionary computation and combinatorial optimization.

EvoCOP	LNCS vol.	Submitted	Accepted	Acceptance (%)
2020	12102	37	14	37.8
2019	11452	37	14	37.8
2018	10782	37	12	32.4
2017	10197	39	16	41.0
2016	9595	44	17	38.6
2015	9026	46	19	41.3
2014	8600	42	20	47.6
2013	7832	50	23	46.0
2012	7245	48	22	45.8
2011	6622	42	22	52.4
2010	6022	69	24	34.8
2009	5482	53	21	39.6
2008	4972	69	24	34.8
2007	4446	81	21	25.9
2006	3906	77	24	31.2
2005	3448	66	24	36.4
2004	3004	86	23	26.7
2003	2611	39	19	48.7
2002	2279	32	18	56.3
2001	2037	31	23	74.2

We would like to express our appreciation to the various persons and institutions making EvoCOP 2020 a successful event. Firstly, we thank the local organization team, led by Francisco Fernández de Vega from the University of Extremadura, Spain, and Federico Divina from the University Pablo de Olavide, Spain. We extend our acknowledgments to Francisco Chicano from the University of Málaga, Spain, and João Correia from the University of Coimbra, Portugal, for the EvoStar website and publicity, as well as Nuno Lourenço for additional general support. Thanks are also due to our EvoStar coordinator Anna I Esparcia-Alcázar, from Universitat Politècnica de València, Spain, and Jennifer Willies, as well as to the SPECIES (Society for the Promotion of Evolutionary Computation in Europe and its Surroundings) executive board, including Marc Schoenauer (President), Anna I Esparcia-Alcázar (Secretary and Vice-President), and Wolfgang Banzhaf (Treasurer). We finally wish to thank our prominent keynote speakers, José Antonio Lozano from the University of the Basque Country, Spain, and Roberto Serra from the University degli Studi di Modena e Reggio Emilia, Italy.

Special thanks also to Christian Blum, Francisco Chicano, Carlos Cotta, Peter Cowling, Jens Gottlieb, Jin-Kao Hao, Jano van Hemert, Bin Hu, Arnaud Liefooghe,

Manuel Lopéz-Ibáñez, Peter Merz, Martin Middendorf, Gabriela Ochoa, and Günther R. Raidl for their hard work and dedication at past editions of EvoCOP, making this one of the reference international events in evolutionary computation and metaheuristics.

April 2020   Luís Paquete
                                                                         Christine Zarges

Organization

EvoCOP 2020 was organized as a part of EvoStar 2020, jointly with EuroGP 2020, EvoMUSART 2020, and EvoApplications 2020.

Organizing Committee

Conference Chairs

Luís Paquete	University of Coimbra, Portugal
Christine Zarges	Aberystwyth University, UK

Local Organization

Francisco Fernández de Vega	University of Extremadura, Spain
Federico Divina	University Pablo de Olavide, Spain

Publicity Chair

João Correia	University of Coimbra, Portugal

EvoCOP Steering Committee

Christian Blum	Artificial Intelligence Research Institute (IIIA-CSIC), Spain
Francisco Chicano	University of Málaga, Spain
Carlos Cotta	University of Málaga, Spain
Peter Cowling	Queen Mary University of London, UK
Jens Gottlieb	SAP AG, Germany
Jin-Kao Hao	University of Angers, France
Jano van Hemert	Optos, UK
Bin Hu	AIT Austrian Institute of Technology, Austria
Arnaud Liefooghe	University of Lille, France
Manuel Lopéz-Ibáñez	The University of Manchester, UK
Peter Merz	Hannover University of Applied Sciences and Arts, Germany
Martin Middendorf	University of Leipzig, Germany
Gabriela Ochoa	University of Stirling, UK
Günther Raidl	Vienna University of Technology, Austria

Society for the Promotion of Evolutionary Computation in Europe and its Surroundings (SPECIES)

Marc Schoenauer (President)
Anna I Esparcia-Alcázar (Secretary and Vice-President)
Wolfgang Banzhaf (Treasurer)
Jennifer Willies (EvoStar coordinator with Anna I Esparcia-Alcázar)

Program Committee

Richard Allmendinger	The University of Manchester, UK
Marco Baioletti	University of Perugia, Italy
Matthieu Basseur	University of Angers, France
Christian Blum	Artificial Intelligence Research Institute (IIIA-CSIC), Spain
Sandy Brownlee	University of Stirling, UK
Maxim Buzdalov	ITMO University, Russia
Arina Buzdalova	ITMO University, Russia
Josu Ceberio	University of the Basque Country, Spain
Marco Chiarandini	University of Southern Denmark, Denmark
Francisco Chicano	University of Málaga, Spain
Carlos Coello Coello	CINVESTAV-IPN, Mexico
Carlos Cotta	University of Málaga, Spain
Bilel Derbel	University of Lille, France
Karl Doerner	Johannes Kepler University Linz, Austria
Benjamin Doerr	LIX-École Polytechnique, France
Carola Doerr	CNRS and Sorbonne University, France
Paola Festa	Universitá di Napoli Federico II, Italy
Carlos M. Fonseca	University of Coimbra, Portugal
Carlos Garcia-Martinez	University of Córdoba, Spain
Adrien Goeffon	University of Angers, France
Andreia Guerreiro	University of Lisbon, Portugal
Jin-Kao Hao	University of Angers, France
Rudová Hana	Masaryk University, Czech Republic
Geir Hasle	SINTEF Digital, Norway
Bin Hu	AIT Austrian Institute of Technology, Austria
Thomas Jansen	Aberystwyth University, UK
Andrzej Jaszkiewicz	Poznań University of Technology, Poland
Ahmed Kheiri	Lancaster University, UK
Mario Koeppen	Kyushu Institute of Technology, Japan
Timo Kötzing	Hasso Plattner Institute, Germany
Frederic Lardeux	University of Angers, France
Rhyd Lewis	Cardiff University, UK
Arnaud Liefooghe	University of Lille, France
Andrei Lissovoi	University of Sheffield, UK

Manuel López-Ibáñez	The University of Manchester, UK
Jose Antonio Lozano	University of the Basque Country, Spain
Gabriel Luque	University of Málaga, Spain
Krzysztof Michalak	Wrocław University of Economics, Poland
Alberto Moraglio	University of Exeter, UK
Christine L. Mumford	Cardiff University, UK
Nysret Musliu	Vienna University of Technology, Austria
Frank Neumann	University of Adelaide, Australia
Gabriela Ochoa	University of Stirling, UK
Pietro Oliveto	University of Sheffield, UK
Beatrice Ombuki-Berman	Brock University, Canada
Luís Paquete	University of Coimbra, Portugal
Mario Pavone	University of Catania, Italy
Paola Pellegrini	French Institute of Science and Technology for Transport, France
Francisco B. Pereira	Polytechnic Institute of Coimbra, Portugal
Daniel Porumbel	CNAM, France
Jakob Puchinger	SystemX-Centrale Supélec, France
Günther Raidl	Vienna University of Technology, Austria
María Cristina Riff	Universidad Técnica Federico Santa María, Chile
Marcus Ritt	Universidade Federal do Rio Grande do Sul, Brazil
Eduardo Rodriguez-Tello	CINVESTAV, Mexico
Andrea Roli	Universitá di Bologna, Italy
Peter Ross	Edinburgh Napier University, UK
Valentino Santucci	University of Perugia, Italy
Frederic Saubion	University of Angers, France
Patrick Siarry	Paris-Est Créteil University, France
Kevin Sim	Edinburgh Napier University, UK
Jim Smith	University of the West of England, UK
Dirk Sudholt	University of Sheffield, UK
Thomas Stützle	Université Libre de Bruxelles, Belgium
El-ghazali Talbi	University of Lille, France
Sara Tari	University of Lille, France
Renato Tinós	University of São Paulo, Brazil
Nadarajen Veerapen	University of Lille, France
Sebastien Verel	Université du Littoral Cote d'Opale, France
Markus Wagner	University of Adelaide, Australia
Darrell Whitley	Colorado State University, USA
Carsten Witt	Technical University of Denmark, Denmark
Takeshi Yamada	NTT Communication Science Laboratories, Japan
Christine Zarges	Aberystwyth University, UK

Contents

Optimizing Prices and Periods in Time-of-use Electricity Tariff Design Using Bilevel Programming

Maria João Alves[1]([⊠]), Carlos Henggeler Antunes[2], and Inês Soares[3]

[1] CeBER and Faculty of Economics, University of Coimbra/INESC Coimbra, Coimbra, Portugal
mjalves@fe.uc.pt
[2] INESC Coimbra, Department of Electrical Engineering and Computers, University of Coimbra, Coimbra, Portugal
ch@deec.uc.pt
[3] INESC Coimbra, Coimbra, Portugal
inesgsoares@gmail.com

Abstract. In this paper, a comparison is made between two bilevel programming models to design time-of-use tariffs in the electricity retail market. The upper-level objective function consists of the maximization of the retailer's profit and the lower-level problem relates to the minimization of the consumer's cost. In the first model, the periods in which prices apply are pre-defined and the aim is to determine the price values. In the second model, which is developed for the first time in this paper, both the periods and prices are decision variables, thus leading to a very large search space for the upper-level problem due to the number of combinations periods-prices. For the model with variable periods, a hybrid approach combining a genetic algorithm for the upper-level search with a mixed-integer linear programming solver to obtain optimal solutions to the lower-level problem is herein developed. Computational results comparing the two models are presented.

Keywords: Bilevel optimization · Genetic algorithm · Mixed-integer linear programming model · Time-of-use pricing · Electricity retail market · Demand response

1 Introduction

Major changes are underway in the electricity sector, namely regarding the evolution to smart grids, the increasing share of renewable sources, the dissemination of electric vehicles, the deployment of distributed storage and the empowerment of consumers/prosumers. Retail electricity markets are very competitive and retail companies should design appropriate pricing schemes to offer to consumers, who are increasingly sensitive to the need to manage consumption patterns in an optimal manner by considering cost and comfort dimensions in their

© Springer Nature Switzerland AG 2020
L. Paquete and C. Zarges (Eds.): EvoCOP 2020, LNCS 12102, pp. 1–17, 2020.
https://doi.org/10.1007/978-3-030-43680-3_1

energy decisions. Energy service companies and grid operators provide automated home energy management systems (HEMS), which manage consumption on the consumer's behalf according to his preferences (e.g., time slots for appliance operation). In general, retailers buy energy in spot markets (e.g., day ahead) or through bilateral contracts. These prices seen by the retailer are increasingly influenced by the grid status and the generation mix required to satisfy demand. So, time-of-use prices have been increasingly adopted, thus fostering a "load follows supply" paradigm in such a way that benefits can be obtained for all players in the energy supply chain (generators, grid operators, retailers and consumers).

There is a hierarchical interplay between a retailer and consumers. The retailer establishes time-of-use prices (which can be valid for a long-term contract, e.g. one year, or be dynamic, e.g. announced one day ahead) to maximize profits. The consumer reacts by managing his consumption to minimize the electricity bill, which can be facilitated by the use of HEMS. Therefore, the design of time-of-use pricing schemes, i.e. specifying variable energy prices and the periods in which they apply, considering the demand response, is of utmost importance for the electricity retail business. This problem has been dealt with bilevel optimization (BLO) models, which are well suited to represent this hierarchical decision setting. The retailer is the *leader*, who decides first by setting prices for given periods, and the consumer is the *follower*, who reacts to these prices by determining the appliance schedule that optimizes his cost function. Although the retailer is the first to play, he must consider the consumer's reaction because it affects the retailer's profit.

In BLO, the (lower-level) follower's optimization problem is nested in the constraints of the (upper-level) leader's problem. BLO models are, in general, very difficult to handle theoretically, methodologically and computationally [1]. Most approaches reported in the literature for designing time-of-use tariffs are devoted to computing the energy prices for pre-defined periods (e.g., along one day for daily cycle prices). This problem has been addressed by several authors, e.g. in [2–7] among others, the last one using a trilevel model.

The problem of designing time-of-use pricing schemes becomes more realistic if, in addition to the price values, the periods in which prices apply are also determined as a result of the optimization. However, this problem becomes much more complex to solve. A general framework of models to optimize prices and periods can be set as follows – models M1 to M4 – being the price values decision variables in all of them; a certain discretization of the planning horizon is considered (e.g., 24 h discretized in 96 units of 15 min each) and the periods to define the prices are ordered sets of *time units* (*t.u.*):

(M1) periods are pre-defined;

(M2) periods are variable, each one consisting of a set of contiguous *t.u.* (i.e., each period is defined by a start *t.u.* and an end *t.u.*), with a pre-specified number of different periods (which implies a maximum number of different prices for the whole planning horizon);

(M3) a price is assigned to each *t.u.*, but imposing a pre-specified maximum number of different prices for the whole planning horizon;

(M4) no restrictions, i.e. each $t.u.$ may have a different price.

Model M4 is the most general one (prices and periods are totally free) and model M1 is the most restricted one (periods are pre-defined). Model M3 is more general than model M2: in M3, prices may change from each $t.u.$ to the next, while in M2 prices may change only by the pre-defined number of periods. In other words, M3 is an extension of M2, looking at each period as a set of possibly non-contiguous $t.u.$

These four time-of-use pricing models may have further variations due to additional constraints on periods (e.g., minimum number of $t.u.$ in each period) or prices (e.g., minimum difference between different prices). Further constraints may be introduced in the time dimension or in price magnitudes.

Models M1 and M2 are more realistic to be implemented in electricity retail markets, namely having in mind consumer's acceptance. By increasing the number of combinations of prices and periods, model M2 offers the retailer the expectation of increasing profits with respect to the ones obtained with model M1 (which is a particular case of M2). This paper is devoted to model M2: a maximum number of different periods is specified; the problem consists of determining the start and the end $t.u.$ for each period and the corresponding price value. A comparison with model M1 is carried out.

Our previous works have been devoted to model M1 [4–6] and M3+M4 [8] using hybrid approaches combining a meta-heuristic (particle swarm optimization or genetic algorithm) to perform the upper-level search for prices and a solver to obtain the solution to the lower-level mixed-integer linear programming (MILP) problem for each price setting. Although models M3 and M4 are interesting from a conceptual perspective, in practice they may induce an excessive change of prices that probably would not be accepted by consumers, or even regulatory authorities, as a viable tariff option.

It should be noticed that the problem of designing time-of-use electricity tariffs falls into the broad category of *price setting problems* [9,10] which includes, e.g., the toll setting problem (the problem of defining highway tolls, where costumers want to minimize their individual generalized travel costs). In these problems, the leader typically seeks to maximize revenues (or profits) raised from taxes or tariffs, while consumers specify consumption or production levels aiming to minimize costs. Therefore, the present study may also act as a lever for future works in pricing setting problems aiming to balance demand along time, which have not yet been addressed in the literature using optimization models. An example may be the definition of happy hours and drink/food prices in bars and restaurants.

In Sect. 2, the main concepts of BLO are presented and bilevel models for price setting problems in the electricity retail market are outlined, considering (i) variable prices only (M1 model) and (ii) variable prices and periods of time (variable period model – M2 model). In Sect. 3, a genetic algorithm for the variable period model is described. Numerical results comparing the two models are presented in Sect. 4 and the main conclusions are drawn in Sect. 5.

2   Bilevel Modelling of Electricity Prices

A general BLO problem can be formulated as follows:

$$\max_{x \in X} \ F(x,y)$$

$$s.t. \ G(x,y) \leq 0$$

$$y \in \arg\min_{y \in Y} \{f(x,y) : g(x,y) \leq 0\}$$

where $x \in \mathbb{R}^{n_1}$ is the vector of variables controlled by the *leader* – the decision maker at the upper-level problem – and $y \in \mathbb{R}^{n_2}$ is the vector of variables controlled by the *follower* – the decision maker at the lower-level problem.

In a bilevel problem, the decision process is sequential as the leader makes his decisions first by setting the values of the variables x. Then, the follower reacts by choosing the y values that optimize his objective function on the feasible solutions restricted by the fixed x. The bilevel problem is the leader's problem. However, the leader must incorporate into the optimization process the reaction of the follower because this affects the leader's objective value and even the feasibility of the solution. It is difficult to find global optimal solutions to bilevel optimization problems due to their inherent non-convexity. Even the linear bilevel problem is NP-hard [1].

In Alves et al. [4], a bilevel problem was considered to model the interaction between the electricity retailer (leader) and a cluster of consumers (follower) with similar consumption and demand response profiles. The retailer buys energy in the wholesale market and wants to determine the prices x_i to be charged to the consumers in I pre-defined periods P_i ($i = 1, \ldots, I$) of a planning horizon discretized into T time units ($t = 1, \ldots, T$), in order to maximize his profit. The consumer aims to minimize the electricity bill, by reacting to the electricity prices communicated by the retailer and deciding on the operation of controllable appliances. In [4] only shiftable appliances were considered, in addition to a base load not deemed for control (e.g., tv set, oven, fridge, etc.). *Shiftable* appliances are typically cyclic loads, such as dishwashers or laundry machines, whose operation cycle can be shifted in time but not interrupted once initiated.

In Soares et al. [6], the bilevel model in [4] was extended by including other types of controllable appliances with different physical features and type of control: in addition to shiftable appliances, a thermostatic load (air conditioning system) and interruptible appliances have been modelled in the lower-level problem. *Interruptible* appliances are loads whose operation can be interrupted provided that the necessary amount of energy is supplied during a required time slot (e.g. charge of an electric vehicle). In both studies [4,6] the lower-level optimization model is a mixed-integer programming problem, which can be solved by an exact MILP solver (for instance, CPLEX) for each instantiation of the upper-level variables x.

In the current study, we consider bilevel models with a consumer's problem including J shiftable appliances ($j = 1 \ldots J$) and K interruptible appliances ($k =$

$1 \ldots K$) as the controllable loads. The consumer wants to determine the times of the operation of these loads in order to minimize the electricity bill, ensuring that the operation of each load is within a specified comfort time slot defined by a start $t.u.$ and an end $t.u.$ within the planning horizon $\{1, \ldots, T\}$: $[t_j^1, t_j^2], j = 1 \ldots J$ and $[t_k^1, t_k^2], k = 1 \ldots K$, respectively, for shiftable and interruptible appliances. Each shiftable appliance j has a load diagram specifying the power (q_{jr}^{shif}) required at each *stage* r (one $t.u.$) of its operation cycle with duration d_j. Each interruptible appliance k has the requirement that the energy E_k should be supplied during the given comfort time slot, being q_k^{int} the power requested at each $t.u.$ when the load is operating. The electricity bill includes an energy component (cost of the energy consumed by all loads) and a power component (the retailer defines multiple levels of contracted power, $P_l^{\text{cont}}, l = 1 \ldots L$, with different prices e_l and the consumer pays for the power level corresponding to the peak).

The electricity prices in each period P_i are controlled at the upper-level: x_i, $i = 1, \ldots, I$, where I is the number of periods P_i.

The lower-level decision variables are binary variables: v_{kt}, which specify, for each interruptible appliance k, whether it is operating or not at each time unit t of the respective comfort time slot; in the case of shiftable appliances, the binary variables are w_{jrt}, which further include the index r to specify the *stage* of the operation cycle in which the load is operating at each t. These binary variables define auxiliary real variables p_t, $\forall t$, which represent the power requested from the grid by all loads: shiftable, interruptible and also a (constant) base load b_t not deemed for control. These variables, together with the electricity prices set by the leader, define the cost of energy for the consumer: $\sum_{i=1}^{I} \sum_{t \in P_i} x_i p_t$. Binary decision variables $u_l \in \{0, 1\}, l = 1 \ldots L$, are also used to model the power component, identifying the peak power level the consumer should be charged for in the whole planning horizon: $\sum_{l=1}^{L} e_l u_l$ (the constraints ensure that only one u_l is equal to 1).

The formulation of the lower-level combinatorial optimization problem is:

$$\min_{p,u} f = \sum_{i=1}^{I} \sum_{t \in P_i} x_i p_t + \sum_{l=1}^{L} e_l u_l \tag{1}$$

s.t.

$$p_t = b_t + \sum_{j=1}^{J} \sum_{r=1}^{d_j} q_{jr}^{\text{shif}} w_{jrt} + \sum_{k=1}^{K} q_k^{\text{int}} v_{kt}, \qquad t = 1, \ldots, T \tag{2}$$

$$\sum_{r=1}^{d_j} w_{jrt} \leq 1, \qquad j = 1, \ldots, J; t = t_j^1, \ldots, t_j^2 \tag{3}$$

$$w_{j(r+1)(t+1)} \geq w_{jrt}, \quad j = 1, \ldots, J; r = 1, \ldots, d_j - 1; t = t_j^1, \ldots, t_j^2 - 1 \tag{4}$$

$$\sum_{t=t_j^1}^{t_j^2} w_{jrt} = 1, \qquad j = 1, \ldots, J; r = 1, \ldots, d_j \tag{5}$$

$$\sum_{t=t_j^1}^{t_j^2-d_j+1} w_{j1t} \geq 1, \qquad j = 1, \ldots, J \tag{6}$$

$$w_{jrt} = 0, \quad j = 1, \ldots, J; r = 1, \ldots, d_j; t < t_j^1 \lor t > t_j^2 \tag{7}$$

$$\sum_{t=t_k^1}^{t_k^2} q_k^{\text{int}} v_{kt} = E_k \qquad k = 1, \ldots, K \tag{8}$$

$$v_{kt} = 0, \quad k = 1, \ldots, K; t < t_k^1 \lor t > t_k^2 \tag{9}$$

$$\sum_{l=1}^{L} u_l = 1 \tag{10}$$

$$p_t \leq \sum_{l=1}^{L} P_l^{\text{cont}} u_l, \qquad t = 1, \ldots, T \tag{11}$$

$$u_l \in \{0,1\}, \quad l = 1, \ldots, L \tag{12}$$

$$w_{jrt} \in \{0,1\}, \quad j = 1, \ldots, J; r = 1, \ldots, d_j; t = t_j^1, \ldots, t_j^2 \tag{13}$$

$$v_{kt} \in \{0,1\}, \quad k = 1, \ldots, K; t = t_k^1, \ldots, t_k^2 \tag{14}$$

where constraints (2) define the power requested at each t by all loads, constraints (3)–(7) model the operation of the shiftable appliances, (8)–(9) model the operation of the interruptible appliances and (10)–(11) model the contracted power.

For a given $x = (x_1, \ldots, x_I)$, the lower-level problem is a MILP problem with a large number of binary variables and constraints. The genetic algorithm used to perform the upper-level search calls the MILP solver CPLEX to solve the lower-level problem (as a black-box).

2.1   Bilevel Model with Pre-defined Periods (M1)

The bilevel model M1, in which only the prices x_i are decision variables for the retailer because the periods P_i are pre-specified (models in [4–6]), can be stated as follows:

$$\max_{x} F = \sum_{i=1}^{I} \sum_{t \in P_i} x_i p_t + \sum_{l=1}^{L} e_l u_l - \sum_{t=1}^{T} \pi_t p_t$$

$$s.t.$$

$$\underline{x} \leq x_i \leq \bar{x}, \quad i = 1, \cdots, I$$

$$\frac{1}{T} \sum_{i=1}^{I} \bar{P}_i x_i \leq x^{AVG}$$

$$(1) - (14)$$

where π_t is the energy price seen by the retailer in the spot market at each $t \in \{1, \ldots, T\}$ and $\bar{P}_i$ denotes the amplitude of P_i, i.e. $\bar{P}_i = P_i^2 - P_i^1 + 1$ where P_i^1, P_i^2 delimit each period P_i, $i \in \{1, \ldots, I\}$ (the start and end $t.u.$, respectively). In order to enforce market competitiveness of retailer prices, the upper-level constraints impose minimum and maximum values on prices in each period P_i and an average price (x^{AVG}) value during the whole planning horizon. In the present study, we consider the same minimum and maximum price values throughout the planning horizon ($\underline{x}$ and $\bar{x}$, respectively).

2.2   Bilevel Model with Variable Periods (M2)

In model M1, the prices to be established in each pre-defined period were the only decision variables. Model M2 builds on model M1 to offer the retailer the possibility of optimizing not just the price values but also the periods in which they apply. This is accomplished by defining a number of periods (which, for instance, may result from regulatory obligations) and determining their optimal start and end $t.u.$ within the planning horizon. The imposition of a number of periods I constrains the maximum number of different price values.

Each period P_i is defined by a start $t.u.$ (P_i^1) and an end $t.u.$ (P_i^2). Thus, the upper-level variables of M2 are: x_i, P_i^1 and P_i^2, $i = 1, \ldots, I$. Two sets of constraints characterize the upper-level problem of Model M2: constraints on prices x (the same as in M1 but with P_i^1 and P_i^2 being decision variables) and constraints that ensure *continuity* of the periods. Only either P_i^1 or P_i^2 need to be considered, because these variables depend on each other $(P_{i+1}^1 = P_i^2 + 1)$, but both are represented in the model below to improve clarity. *Continuity* constraints ensure that the first period starts at $t = 1$, the last one ends at $t = T$, and the periods are chained: $P_1^1 \leq P_1^2 = P_2^1 - 1, \ldots, P_{I-1}^2 + 1 = P_I^1 \leq P_I^2 = T$. For instance, suppose that one $t.u.$ is 15 min and the planning horizon starts at 00:00 h; $P_1^1 = 1$, meaning that period P_1 includes, at least, the first 15 min of the day; if, for instance, $P_1^2 = 4$, then the first period is [00:00, 01:00[h and the second period starts at $P_2^1 = 5$, which means that P_2 includes at least $t = 5$, i.e., [01:00, 01:15[h. The model M2 can be stated as follows:

$$\max_{x, P^1, P^2} F = \sum_{i=1}^{I} \sum_{t \in [P_i^1, P_i^2]} x_i p_t + \sum_{l=1}^{L} e_l u_l - \sum_{t=1}^{T} \pi_t p_t$$

s.t.

$$\underline{x} \leq x_i \leq \bar{x}, \quad i = 1, \cdots, I$$

$$\frac{1}{T} \sum_{i=1}^{I} (P_i^2 - P_i^1 + 1) x_i \leq x^{AVG}$$

$$P_1^1 = 1; \quad P_I^2 = T$$

$$P_{i+1}^1 = P_i^2 + 1, \quad i = 1, \cdots, I-1$$

$$P_i^2 \geq P_i^1, \quad i = 1, \cdots, I$$

$$P_i^1, P_i^2 \quad \text{integer}, \quad i = 1, \cdots, I$$

$$\min_{p,u} f = \sum_{i=1}^{I} \sum_{t \in [P_i^1, P_i^2]} x_i p_t + \sum_{l=1}^{L} e_l u_l$$

$$\text{s.t.} \quad (2) - (14)$$

Other conditions may be imposed on the periods, which we call *aggregation* constraints, e.g. the following ones.

- The length of each period must be a multiple of a given number of units of time. Consider, for instance, that the planning horizon is discretized in units of quarter-hour (which is generally used for measurements in power systems and enables a fine grain analysis of appliance operation) and the prices must be defined for periods that are multiple of half-hour or one hour (e.g., for regulatory reasons). Therefore, the amplitude of each period $(P_i^2 - P_i^1 + 1)$ must be multiple of 2 or 4, respectively. For a multiple of β units of time, and provided that T is multiple of β, the following constraints are included:

$$P_i^2 - P_i^1 + 1 = \beta k_i, \quad i = 1, \cdots, I$$
$$k_i \text{ integer}, 1 \leq k_i \leq \tfrac{T}{\beta}, \quad i = 1, \cdots, I \qquad (15)$$

- Each period must have a minimum length C, where C is a constant:

$$P_i^2 - P_i^1 + 1 \geq C, \quad i = 1, \cdots, I \qquad (16)$$

In the present study, we have considered constraints (15) with $\beta = 2$, which also ensure a minimum length of $C = 2$. Since the quarter-hour is the *t.u.* considered, the periods are then multiple of half-hour.

The electricity prices charged to the consumer $(x_i, i = 1, \ldots, I)$ and the prices seen by the retailer in the spot market $(\pi_t, t = 1, \ldots, T)$ are presented in €/KWh. So, an adequate scale factor α is applied in the upper and lower-level objective functions of M1 and M2 to convert the prices into the *t.u.* used in these models. That is, x_i and π_t are replaced by αx_i and $\alpha \pi_t$, respectively. In this study, $\alpha = 1/4$.

3   A Genetic Algorithm for the Variable Period Model

We aim at comparing results obtained for models M1 and M2. The algorithm presented in [4] has been adapted to deal with model M1 considering prices with 4 decimal places (as it is usual in electricity bills presented to consumers) instead of real numbers. The algorithm consists of a genetic algorithm (GA) to deal with the upper-level search combined with CPLEX to find the optimal solution to the lower-level problem for each x vector. The individuals dealt with by the GA are the price vectors $x = (x_1, x_2, \ldots, x_I)$. The upper-level constraints of M1 are ensured by a *repair* routine [4], which has been adjusted in the present work to prices with a fixed number of decimal places and is also used in the approach developed for model M2. This routine is briefly described below.

The approach developed for the variable period model M2 is also a hybrid GA-solver. The individuals are composed by two vectors, one for prices (as in model M1) and the other for periods. Since the start $t.u.$ of each period is the end $t.u.$ of the previous period $+1$ ($P_i^1 = P_{i-1}^2 + 1$), the periods are represented only by their end $t.u.$: $P^2 = (P_1^2, P_2^2, \ldots, P_{I-1}^2, T)$ with increasing integer values $P_i^2 < P_{i+1}^2$. This vector has dimension I, but the last component is fixed to T. These two vectors define each upper-level solution and are illustrated in Fig. 1.

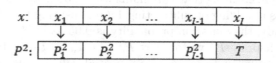

Fig. 1. Encoding of an upper-level solution in model M2

A population of N individuals $(x^n, P^{2,n})$, $n = 1, \ldots, N$ evolves throughout G iterations of the GA. The vector of start $t.u.$ $P^{1,n}$ corresponding to $P^{2,n}$ is: $P^{1,n} = (1, P_1^{2,n} + 1, \ldots, P_{I-1}^{2,n} + 1)$. For each individual, the lower-level problem of M2 with $P^1 = P^{1,n}$, $P^2 = P^{2,n}$ and $x = x^n$ is exactly solved. Let $y^n = (p^n, u^n)$ be the optimal solution obtained for this lower-level instance (vector p of power required by all load operation and vector u of binary variables that determine the contracted power). Each solution is then evaluated by the upper-level objective function $F(x^n, P^{1,n}, P^{2,n}, y^n)$, which gives its *fitness*.

The general description of the GA is presented below.

Step 1 - Create the initial population with N individuals ($n = 1, \cdots, N$)
  - Generate each $P^{2,n}$: $I - 1$ different integer numbers are randomly drawn; these values are then sorted by increasing order. In order to satisfy the time limits and the *aggregation* constraints (15), multiples of β are generated in the range $(1, T - 1)$; $P_I^{2,n} = T, \forall n$.
  - Generate each x^n: a real number with 4 decimal places is randomly generated in $[\underline{x}, \bar{x}]$ for each component x_i^n; x^n is then repaired to satisfy also the average price constraint (*repair* routine).
Step 2 - Obtain the lower-level solutions: for each individual, the $P^{1,n}$ vector associated with $P^{2,n}$ is defined and the lower-level problem is solved for $(x^n, P^{1,n}, P^{2,n})$ to obtain y^n. Compute its fitness value, F^n.
While the maximum number of generations G is not achieved **do**
Step 3 - Selection and Reproduction
  - Select N pairs of individuals for being parents: one parent is randomly chosen and the other is the winner of a binary tournament with replacement.

- In order to ensure that all P^2 values will satisfy the *aggregation* constraints, their scale is changed by dividing all values by β, which will be recovered at the final of the reproduction phase; after scaling, P_I^2 $= \tilde{T} = T/\beta$.
- For each pair of parents $(P^{2'}, x')$ and $(P^{2''}, x'')$, an one-point crossover operator is separately applied to P^2 and x. For P^2, the last component (equal to $\tilde{T}$) never changes.
- Apply mutation to each offspring with a given probability of changing each gene of x and of P^2. The mutation of an x_i consists of adding or subtracting a positive perturbation randomly generated in the range $[0, 0.2(\bar{x}-\underline{x})]$, ensuring that x_i remains within bounds. The mutation of P_i^2 consists of adding or subtracting 1 $t.u.$, ensuring that $1 \leq P_i^2 < \tilde{T}$. In this study, the mutation probability is 0.05 as in [4].
- For each offspring P^2 vector, sort P_i^2 $(i = 1, \ldots, I - 1)$ by increasing order and adjust it to contain no repeated values, since there are no periods with duration 0; convert P^2 to the original scale (see Fig. 2 for an example).
- Repair each x vector of the offspring to also satisfy the average price constraint and keeping 4 decimal places (*repair* routine).
- Obtain the lower-level solutions for the offspring as in *Step 2*.

Step 4 - Replacement
Form the next population by copying the solution with best F obtained thus far (which is either in the current population or in the offspring) and by performing $N - 1$ binary tournaments without replacement between individuals of the current population and the offspring population.
End While
Return the solution with the highest fitness F.

Figure 2 illustrates the reproduction process (without mutation) of two 6-period vectors, $P^{2'}$ and $P^{2''}$, for a planning horizon of $T = 96$ $t.u.$ of quarter-hour and $\beta = 2$ (periods should be multiple of half-hours). The vectors $P^{2'}$ and $P^{2''}$ at the top of Fig. 2 have already been scaled to half-hour units (thus, $\tilde{T} = 48$). After applying crossover and sorting by ascending order the values of the offspring P^2 $(P_i^2, i = 1, \ldots, I - 1)$, if there are two equal values then the second one is incremented by 1. In this example, this only happened once for value 27. If a modified value becomes equal to the next, then this process is repeated until all duplicate values are eliminated. If the second to the last value (P_{I-1}^2) is equal to the last one $(\tilde{T})$, the chromosome is discarded and another one must be generated.

In this work, a one-point crossover operator has been used both for P^2 and x vectors. A geometric crossover operator for the prices x had been used in the study [8] for models M3 and M4, but the results with the GA were not encouraging.

The *repair* routine implemented in both algorithms (for M1 and M2 models) ensures that prices are within bounds ($\underline{x} \leq x_i \leq \bar{x}, i = 1, \ldots, I$) and satisfy the average price constraint ($\frac{1}{T} \sum_{i=1}^{I} \bar{P}_i x_i \leq x^{AVG}$ with $\bar{P}_i = P_i^2 - P_i^1 + 1$). Since

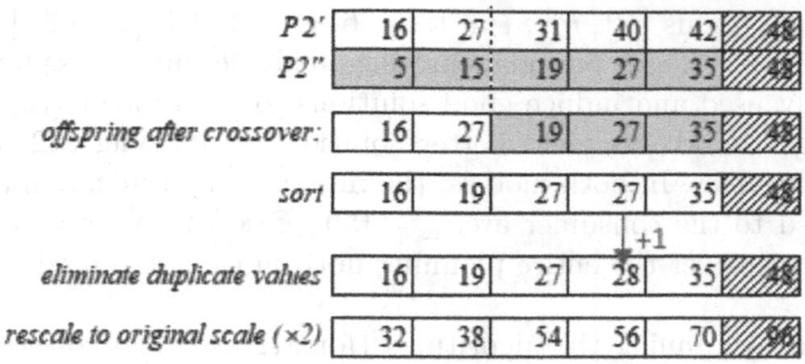

P2'	16	27	31	40	42	48
P2"	5	15	19	27	35	48
offspring after crossover:	16	27	19	27	35	48
sort	16	19	27	27	35	48
				+1		
eliminate duplicate values	16	19	27	28	35	48
rescale to original scale (×2)	32	38	54	56	70	96

Fig. 2. Crossover and adjustment of periods

the aim is to maximize the retailer's profit, the repair operations attempt to set prices as close as possible to satisfying the average price constraint as an equality, keeping all values with a fixed number of d decimal places (we have been working with $d = 4$). The *repair* routine operates as follows. Firstly, the x_i are truncated to d decimal places and pushed into bounds. Let A be the set of indices i of x that can still change; initially, $A = \{1, \ldots, I\}$. The following cycle is repeated until a valid x is obtained or $A = \emptyset$:

(1) $x_i \leftarrow trunc(x_i + \delta, d)$, $\forall i \in A$, with

$$\delta = \frac{Tx^{AVG} - \sum_{i=1}^{I} \bar{P}_i x_i}{\sum_{i \in A} \bar{P}_i}$$

(2) if all x_i are within the bounds, then stop: a valid x has been obtained; otherwise, push into the closest bound ($\underline{x}$ or $\bar{x}$) any x_i that is out of bounds, $A \leftarrow A \setminus \{i\}$ and return to (1) if $A \neq \emptyset$.

4   Results

A case study has been used to compare results obtained with the M1 and M2 models using the respective hybrid GA algorithms. A 24-h planning horizon (starting at 00:00 h) is considered, which is discretized into *t.u.* of quarter-hour, leading to a planning horizon $\{1, \ldots, 96\}$. The consumer's problem includes five controllable appliances: three shiftable loads (dishwasher, laundry machine and clothes dryer) and two interruptible loads (electric water heater and electric vehicle). The data concerning the consumer's problem were obtained from actual audit information and some values were estimated; they can be found in the Supplementary Material of [6], including the operation cycles of the loads, the comfort time slots allowed for the operation of each load, the base load, the contracted power levels and their costs, as well as the prices seen by the retailer at the spot market. These data define a lower-level problem with 559 binary variables.

Six periods of time $P_i, i = 1, \ldots, 6$, are considered for defining the electricity prices to be charged by the retailer to the consumer (as in [6]). In model M1, the pre-defined periods $[P_i^1, P_i^2]$, $i = 1, \ldots, 6$, are: [1–28], [29–44], [45–56], [57–72], [73–84], [85–96]. These periods reproduce realistic time-of-use tariff schemes being currently used and induce good solutions to the retailer (thus, imposing more challenges to M2 to yield better solutions). In model M2, the periods are decision variables. In both models, the minimum and maximum prices that can be charged to the consumer are: $\underline{x} = 0.08$ €/kWh and $\bar{x} = 0.35$ €/kWh. The average price over the entire planning horizon cannot exceed $x^{AVG} = 0.18$ €/kWh.

We started by running the algorithms (for M1 and M2) considering different parameterizations of the population size (N) and number of generations (G) in order to choose a configuration that provides a satisfactory compromise between the quality of the solution obtained vs. the computation time needed. The following parameterizations were attempted for $N \times G$: 30×100, 30×200, 40×150, the last two requiring similar computation effort, since both require solving 6000 lower-level MILP problems, which is the most demanding part of the computational effort. Ten independent runs were performed with M1 and M2 for each parameterization. In order to allow a better comparison, an equal *rand-seed* (seed used for the generation of random numbers) was considered for runs with the same index. For instance, run k started with *rand-seed* r_k for all parameterizations $N \times G$ in M1 and M2. The best, worst, average and standard deviation of the retailer's profit (F) obtained over the 10 runs for each parametrization in each model are reported in Table 1. All F values are in € and refer to a cluster of 1000 consumers with similar consumption and demand response profiles. The best values for each model are highlighted in bold.

Table 1. Results of different combinations of the population size and number of generations for M1 and M2

F values	M1 ($N \times G$)			M2 ($N \times G$)		
	30×100	30×200	40×150	30×100	30×200	40×150
Maximum	6038.67	**6039.24**	**6039.27**	6145.12	**6151.16**	6145.75
Minimum	6026.96	**6035.99**	6028.68	**6070.72**	**6070.72**	**6070.72**
Average	6034.49	**6037.57**	6036.80	6116.18	**6121.14**	6108.95
Stand.dev.	3.387	0.963	2.773	26.930	27.813	21.981

From this experiment, we can observe (Table 1):

- Better retailer's profit can be obtained when the retailer can set prices and periods (model M2) over setting prices only (model M1) (1.85% improvement in the best cases). This result was expected, because the M2 solution space, say $S(M2)$, includes the M1 solution space, i.e., $S(M1) \subseteq S(M2)$. Although theoretically expected, it should be noticed that in our previous work [8] with models with more degrees of freedom – referred to as M3 and M4 above – the population-based approaches experienced several difficulties to efficiently explore broader upper-level spaces, yielding better results for the M3 model than for M4, in spite of M3 being a constrained M4. In the present work this has not happened: the algorithm proposed herein provides good results for M2, with values for F systematically higher than the ones obtained with M1, also providing better figures (in any of the parameterizations) than the results obtained in [8] for M3 and M4, although $S(\text{M2}) \subseteq S(\text{M3}) \subseteq S(\text{M4})$.

- The improvement of the results from 30×100 to 30×200 is small, being slightly higher in M2 than in M1. The average improvement in M1 is 0.05%, while in M2 is 0.08%. The solution space $S(\text{M2})$ is much larger than $S(\text{M1})$ due to the combinatorial explosion of the price-periods combinations. This may explain why M2 benefits more than M1 from a longer search process; this may also justify the higher standard deviations in M2 than in M1. The improvement of F from 100 to 200 generations does not seem relevant given the large increase in the computational effort, which doubles.
- The parameterization 40×150 does not produce better results, still looking slightly worse than 30×200, although 10 runs are not enough to support strong conclusions.

The non-parametric Kruskal-Wallis test was applied to assess whether the differences of the F values obtained with the different parameterizations in each model are statistically significant, considering a significance level of $\alpha = 0.01$. In both models, the differences are not statistically significant. The Mann-Whitney test comparing the results of M1 and M2 for the same parameterizations led to the conclusion that the differences are statistically significant in all the three cases.

The algorithms were run in a computer with an Intel Core i7-7700 CPU 3.6 GHz, 64 GB RAM. The computation time of each generation is similar for M1 and M2, which is on average less than 4" for the population size of 30 and 5" for the population size of 40. The total computation time of one complete run is about 6'– 6'30" for the parametrization 30×100 and about 12' – 13'30" for the parametrizations 30×200 and 40×150, which have similar computation times.

Given the results obtained in this experiment, we have adopted the 30×100 parameterization because it presents a good compromise between solution quality and computation time. The best solution (maximum F) obtained for M1 has $F = 6038.67$ and the best solution obtained for M2 has $F = 6145.12$. We refer to these solutions as Sol_{M1} and Sol_{M2}, respectively. The periods $[P_i^1, P_i^2]$, $i = 1, \ldots, 6$, computed in Sol_{M2} are: [1–8], [9–12], [13–16], [17–48], [49–52], [53–96]. Figure 3 compares the pre-defined periods in M1 with the ones obtained with M2, showing the times of the day (h) that delimit the periods; $\{1, \ldots, 96\}$ corresponds to 00:00 h – 24:00 h where $t = 1$ represents the $t.u.$ from 00:00 h to 00:15 h, and so on. A significant difference between the periods of M1 and M2 can be observed: P_1 is the longer period in M1 ranging from 00:00 h to 07:00 h, while M2 defines three of the six periods from 00:00 h to 04:00 h. The longer period of M2 ranges from 13:00 h to 24:00 h.

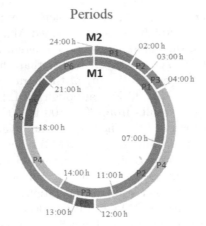

Fig. 3. Pre-defined periods in M1 and the best periods given by the algorithm for M2 in the 30×100 parameterization (Sol_{M2})

Considering the periods of M2 in Sol_{M2} (shown in Fig. 3), we have further intensified the search in an attempt to improve the prices for these periods. Accordingly, these periods were fixed, and the M1 model was solved for these pre-defined periods. Five independent runs were performed from scratch (without injecting Sol_{M2} in the initial population) for the parameterization $N \times G = 30 \times 100$. The solutions obtained ranged from $F = 6147.0$ to $F = 6155.5$, with an average of $F = 6151.1$, i.e., all improving $F(\text{Sol}_{M2}) = 6145.12$. The best solution ($F = 6155.5$) is also better that the one provided by running M2 for 30×200. This result suggests that it is better to execute the algorithm for the variable period model M2 during 100 iterations and then intensify the search for prices by using M1 for the best periods obtained than extending the search for periods and prices using M2 until 200 iterations. We have further experimented to run M2 for $N \times G = 30 \times 500$. The best and average values obtained over 10 runs were $F = 6152.85$ and $F = 6126.82$, both being worse than the respective values obtained with the strategy M2+M1 (with $G = 100$ in each one).

Let us denote by Sol_{M2+M1} the best solution obtained in the experiment M2+M1, which has the same periods as Sol_{M2} but slightly different prices. Table 2 shows the prices in Sol_{M1}, Sol_{M2} and Sol_{M2+M1}. Figure 4 compares the best prices obtained for M1 (Sol_{M1} with a retailer's profit of 6038.67) and for M2 (Sol_{M2+M1} with a retailer's profit of 6155.50). Although a maximum of 6 different prices was allowed, solutions Sol_{M1} and Sol_{M2+M1} have fewer than 6 different prices (4 and 5, respectively).

Table 2. Prices in the best solutions obtained for M1 and M2 in the 30×100 parameterization (Sol_{M1} and Sol_{M2}) and after the intensification of the search for M2 with M1 (Sol_{M2+M1})

		Prices (€/kWh)					
		P_1	P_2	P_3	P_4	P_5	P_6
	Periods	[1–28]	[29–44]	[45–56]	[57–72]	[73–84]	[85–96]
Sol_{M1}	Prices	0.2976	0.2986	0.08	0.0803	0.0803	0.08
	Periods	[1–8]	[9–12]	[13–16]	[17–48]	[49–52]	[53–96]
Sol_{M2}	Prices	0.2689	0.2384	0.2701	0.2890	0.0804	0.0801
Sol_{M2+M1}	Prices	0.2683	0.2360	0.2793	0.2885	0.08	0.08

Fig. 4. Prices obtained for the models M1 (Sol_{M1}) and M2 (Sol_{M2+M1}).

5 Conclusions

This paper presented a comparison between two bilevel programming models to assist electricity retail companies to design optimal time-of-use tariffs. In the upper-level problem, the retailer maximizes the profit and in the lower-level problem the consumer minimizes the cost using his flexibility in the use of appliances in face of the time-differentiated prices. A bilevel model was previously presented by the authors in which the periods for setting the different prices were pre-defined and the aim was to determine the price values that maximize the retailer's profit. In this paper, a new more general model is proposed in which both the periods and prices are decision variables, thus leading to a very large search space for the upper-level problem due to the vast number of combinations periods-prices. To deal with this variable period model, a hybrid approach combining a genetic algorithm for the upper-level search with a mixed-integer linear programming solver to obtain optimal solutions to the lower-level problem has been developed. Specific encoding as well as crossover and mutation operators have been designed to make the most of the physical features of the problem.

The algorithm has been able to compute good quality solutions obtaining higher profit when the retailer can establish prices and periods over setting prices only, with a moderate computation effort. This information is of utmost importance for a retailer designing tariff options to offer consumers in very competitive electricity retail markets.

In order to cope with the complexity of the variable period model, additional experiments consisted of using the period configuration determined in this new model as an input of the pre-defined period model aiming to further improve prices. This intensification strategy proved useful since better solutions have been obtained in comparison with solutions found with a higher computation effort in the variable period model.

Further work will involve a comprehensive study of time-of-use pricing problems vs. adequate features of algorithmic approaches, including using algorithms based on strategies other that the hybridization of metaheuristics with mathematical programming solvers.

Acknowledgments. This work was partially supported by projects UIDB/00308/ 2020 and by the European Regional Development Fund through the COMPETE 2020 Programme, FCT - Portuguese Foundation for Science and Technology and Regional Operational Program of the Center Region (CENTRO2020) within projects ESGRIDS (POCI-01-0145- FEDER-016434) and MAnAGER (POCI- 01-0145-FEDER-028040).

References

1. Dempe, S.: Foundations of Bilevel Programming. Springer, Heidelberg (2002). https://doi.org/10.1007/b101970
2. Zugno, M., Morales, J.M., Pinson, P., Madsen, H.: A bilevel model for electricity retailers' participation in a demand response market environment. Energy Econ. **36**, 182–197 (2013)
3. Meng, F.L., Zeng, X.J.: A bilevel optimization approach to demand response management for the smart grid. In: 2016 IEEE Congress on Evolutionary Computation (CEC), pp. 287–294 (2016)
4. Alves, M.J., Antunes, C.H., Carrasqueira, P.: A hybrid genetic algorithm for the interaction of electricity retailers with demand response. In: Squillero, G., Burelli, P. (eds.) EvoApplications 2016. LNCS, vol. 9597, pp. 459–474. Springer, Cham (2016). https://doi.org/10.1007/978-3-319-31204-0_30
5. Carrasqueira, P., Alves, M.J., Antunes, C.H.: Bi-level particle swarm optimization and evolutionary algorithm approaches for residential demand response with different user profiles. Inf. Sci. (Ny) **418–419**, 405–420 (2017)
6. Soares, I., Alves, M.J., Antunes, C.H.: Designing time-of-use tariffs in electricity retail markets using a bi-level model - computing bounds when the lower level problem cannot be exactly solved. Omega (2019, in press)
7. Aussel, D., Brotcorne, L., Lepaul, S., von Niederhäusern, L.: A trilevel model for best response in energy demand-side management. Eur. J. Oper. Res. **281**, 299–315 (2020)
8. Soares, I., Alves, M.J., Antunes, C.H.: A bi-level optimization approach to define dynamic tariffs with variable prices and periods in the electricity retail market. In: Proceedings of the EUROGEN 2019, Springer ECCOMAS book series on Computational Methods in Applied Sciences (2020, accepted for publication)

9. Marcotte, P., Savard, G.: A bilevel programming approach to optimal price setting. In: Zaccour, G. (ed.) Decision & Control in Management Science: Essays in Honor of Alain Haurie, vol. 4, pp. 97–117. Springer, Boston (2002). https://doi.org/10.1007/978-1-4757-3561-1_6
10. Labbé, M., Violin, A.: Bilevel programming and price setting problems. 4OR **11**, 1–30 (2013)

An Algebraic Approach for the Search Space of Permutations with Repetition

Marco Baioletti[1] (ID), Alfredo Milani[1] (ID), and Valentino Santucci[2]([⊠]) (ID)

[1] Department of Mathematics and Computer Science, University of Perugia,
Perugia, Italy
{marco.baioletti,alfredo.milani}@unipg.it
[2] Department of Humanities and Social Sciences,
University for Foreigners of Perugia, Perugia, Italy
valentino.santucci@unistrapg.it

Abstract. We present an algebraic approach for dealing with combinatorial optimization problems based on permutations with repetition. The approach is an extension of an algebraic framework defined for combinatorial search spaces which can be represented by a group (in the algebraic sense). Since permutations with repetition does not have the group structure, in this work we derive some definitions and we devise discrete operators that allow to design algebraic evolutionary algorithms whose search behavior is in line with the algebraic framework. In particular, a discrete Differential Evolution algorithm which directly works on the space of permutations with repetition is defined and analyzed. As a case of study, an implementation of this algorithm is provided for the Job Shop Scheduling Problem. Experiments have been held on commonly adopted benchmark suites, and they show that the proposed approach obtains competitive results compared to the known optimal objective values.

Keywords: Discrete evolutionary algorithms · Permutations with Repetition · Algebraic approach

1 Introduction

An algebraic framework for combinatorial optimization problems has been previously proposed in a series of articles [4–6,22]. This framework mainly proposed the discrete operations of sum, difference, and scalar multiplication that allow to design discrete variants of widely used continuous evolutionary algorithms such as the Differential Evolution (DE) [22] and the Particle Swarm Optimization (PSO) [23]. The main requirement of the framework is that the solutions in the search space of the combinatorial problem at hand must form a group (in the algebraic sense). For instance, this is the case of widely considered search spaces such as those of bit-strings [24] and permutations [1,6].

However, there are interesting problems defined in combinatorial search spaces which do not form a group. One of these is the space of permutations

© Springer Nature Switzerland AG 2020
L. Paquete and C. Zarges (Eds.): EvoCOP 2020, LNCS 12102, pp. 18–34, 2020.
https://doi.org/10.1007/978-3-030-43680-3_2

with repetition, i.e., ordering of items which – differently from classical permutations – can appear multiple times in the sequence. This search space has been considered, for example, in [8,14,16,20,27]. The most notable applications of permutations with repetition are in some scheduling and partitioning problems. Indeed, in the scheduling case, the repeated items accommodate the fact that some jobs need to be processed in more than one machine, while, in partitioning problems, permutations with repetition are intended as assignments of items to a particular cluster among a set of clusters with a given size. Widely known example of such problems are the job shop scheduling problem [11] and the balanced multiway graph partitioning problem [17].

In this work, we extend the algebraic framework in order to work on the search space of permutations with repetition, even if they do not form a group. With this regard, we derive formal definitions and algorithmic implementations of the discrete operators of sum, difference and scalar multiplication. Such operators allow to design discrete variants of evolutionary algorithms which are commonly and effectively used in the continuous search spaces. In particular, we introduce a discrete Algebraic Differential Evolution for Permutations with Repetition (ADE-PR) by also analyzing its search behavior.

ADE-PR can in principle be applied to any problems requiring a permutation with repetition as a solution. As a case of study, we have investigated the effectiveness of ADE-PR on the Job Shop Scheduling Problem (JSSP). Therefore, few additional algorithmic components, purposely defined for the JSSP, have been integrated in ADE-PR. Finally, computational experiments have been held by considering widely used benchmark suites for the JSSP.

The rest of the paper is organized as follows. Section 2 recalls the algebraic framework which has been extended in Sect. 3 in order to handle the space of permutations with repetition. Section 4 describes the main scheme of ADE-PR, while its implementation for the JSSP is depicted in Sect. 5. The experimental analysis is described in Sect. 6, while Sect. 7 concludes the paper by also providing future lines of research.

2   Algebraic Background

2.1   The Abstract Algebraic Framework for Evolutionary Computation

The algebraic framework for evolutionary computation, firstly proposed in [22] and further studied in [3,4,6,7], allows to define the discrete operators $\oplus$, $\ominus$ and $\odot$ which simulate in a discrete search space the properties of their numerical counterparts.

In particular, these discrete operators are abstractly defined for any combinatorial search space whose solution set can be represented with an algebraic structure known as *finitely generated group* [18].

The triple $(X, \circ, G)$ is a finitely generated group representing the search space of a given combinatorial optimization problem $\mathcal{P}$ if:

- X is the set of solutions of $\mathcal{P}$;
- $\circ$ is a binary operation on X satisfying the group properties, i.e., closure, associativity, identity (e), and invertibility (x^{-1}); and
- $G \subseteq X$ is a finite generating set of the group, i.e., any $x \in X$ has a (not necessarily unique) minimal-length decomposition $\langle g_1, \ldots, g_l \rangle$, with $g_i \in G$ for all $i \in \{1, \ldots, l\}$, and whose evaluation is x, i.e., $x = g_1 \circ \cdots \circ g_l$.

Moreover, the length of a minimal decomposition of a discrete solution $x \in X$ is denoted by $|x|$.

Using $(X, \circ, G)$, it is possible to provide formal definitions of the operators $\oplus$, $\ominus$ and $\odot$. Let $x, y \in X$ and $\langle g_1, \ldots, g_k, \ldots, g_{|x|} \rangle$ be a minimal decomposition of x, then

$$x \oplus y := x \circ y, \tag{1}$$

$$x \ominus y := y^{-1} \circ x, \tag{2}$$

$$a \odot x := g_1 \circ \cdots \circ g_k, \text{ with } k = \lceil a \cdot |x| \rceil \text{ and } a \in [0, 1]. \tag{3}$$

Interesting graph-based interpretations of these definitions can be given as follows. The algebraic structure on the search space naturally defines neighborhood relations among the solutions. Indeed, any finitely generated group $(X, \circ, G)$ is associated to a labelled digraph $\mathcal{G}$ whose vertices are the solutions in X and two generic solutions $x, y \in X$ are linked by an arc labelled by $g \in G$ if and only if $y = x \circ g$. Therefore, a simple one-step move in the search space can be directly encoded by a generator, while a composite move can be synthesized as the evaluation of a sequence of generators (a path on the graph).

In analogy with $\mathbb{R}^n$, the elements of X can be dichotomously interpreted both as solutions (vertices on the graph) and as displacements between solutions (labelled paths on the graph). As detailed in [22], this allows to provide rational interpretations of the definitions (1), (2) and (3) as follows:

- $x \oplus y$ is the vertex of $\mathcal{G}$ where we arrive if we move from the vertex x following the arcs in any (minimal) decomposition of y;
- a minimal decomposition of $x \ominus y$ corresponds to the sequence of arcs in a shortest path from the vertex y to the vertex x in $\mathcal{G}$;
- the scalar multiplication $a \odot x$, with $a \in [0, 1]$, corresponds to truncating a shortest path from the vertex e (the identity of the group) to the vertex x in $\mathcal{G}$.

Clearly, these geometrical interpretations are in line with the vectors/points interpretations of the classical Euclidean space.

2.2   The Algebraic Differential Evolution

As shown in [22] and [23], expressions which involve the three discrete operators allow to derive discrete variants of some popular evolutionary schemes originally defined for continuous problems [19, 26]. For instance, a discrete variant of the Differential Evolution (DE) algorithm, namely Algebraic Differential Evolution

(ADE), can be defined by simply replacing the classical mathematical operations with their discrete variants $\oplus, \ominus, \odot$ in the definition of the differential mutation which is the key operator of the DE.

Therefore, the differential mutation of ADE is defined as follows:

$$v \leftarrow x_{r_0} \oplus F \odot (x_{r_1} \ominus x_{r_2}), \tag{4}$$

where $x_{r_0}, x_{r_1}, x_{r_2} \in X$ are three randomly selected population individuals, $F \in [0, 1]$ is the DE scale factor parameter, and $v \in X$ is the mutant produced.

The interpretation of Eq. (4) in the search space graph $\mathcal{G}$ is as follows: v is generated by starting from the vertex x_{r_0} and following the arcs indicated by $F \odot (x_{r_1} \ominus x_{r_2})$ which is a sequence of arcs' labels (generators) obtained by truncating a shortest path from x_{r_2} to x_{r_1}. This is in line with what is done by the classical differential mutation equation in the Euclidean space, i.e., generate a mutant v by applying to x_{r_0} the vector corresponding to the truncated segment which connects x_{r_2} to x_{r_1}. Indeed, note that the concept of segment in the Euclidean space is analogous to the concept of shortest path in the graph representing a discrete search space.

2.3 The Search Space of Permutations

The definitions provided in the previous sections are abstract and require implementations for concrete spaces. One of the most investigated search space that verifies the properties of finitely generated groups is the space of permutations [2,25].

The permutations of the set $\{1, \ldots, n\}$, together with the usual permutation composition, form the so-called *Symmetric group* $\mathcal{S}(n)$. The identity permutation is $\iota = \langle 1, \ldots, n \rangle$. Furthermore, since $\mathcal{S}(n)$ is finite, it is also finitely generated.

One of the most useful generating sets for the permutations is the set of *simple transpositions* $ASW \subset \mathcal{S}(n)$, i.e., particular permutations which algebraically encode the *adjacent swap moves*. Formally,

$$ASW = \{\sigma_i : 1 \leq i < n\}, \tag{5}$$

where the $n - 1$ simple transpositions σ_i are permutations such that

$$\sigma_i(j) = \begin{cases} i + 1 & \text{if } j = i, \\ i & \text{if } j = i + 1, \\ j & \text{otherwise.} \end{cases} \tag{6}$$

Given a generic $\pi \in \mathcal{S}(n)$, the composition $\pi \circ \sigma_i$ swaps the i-th and $(i + 1)$-th items in π. Therefore, using the abstract definitions provided before, a minimal decomposition of the difference between two generic permutations π and ρ corresponds to the shortest sequence of adjacent swap moves which transforms π into ρ.

A minimal decomposition for a generic permutation $\pi \in \mathcal{S}(n)$, in terms of ASW, can be obtained by ordering the items in π by using a sorting algorithm based on adjacent swap moves. The sequence of generators corresponding to the moves performed during the sorting process is annotated, then reversing this sequence produces a minimal decomposition [22].

As widely known, the bubble-sort algorithm sorts any given array by using a minimal number of adjacent swap moves, therefore it can be used for computing a minimal decomposition of any permutation in terms of ASW. Anyway, since there can be more than one minimal decompositions, a randomized variant of bubble-sort, namely $RandBS$, has been proposed in [22].

$RandBS$ exploits the concept of inversion and the property that the identity permutation ι is the only permutation without inversions. Formally, (i, j) is an inversion of a given permutation π if and only if $i < j$ and $\pi(i) > \pi(j)$. Moreover, a permutation with a positive number of inversions has to have at least one *adjacent inversion*, i.e., an inversion of the form $(i, i+1)$. Therefore, $RandBS(\pi)$ decreases the inversions of π by first computing its adjacent inversions and, then, iteratively applying adjacent swaps corresponding to those inversions. At the end of this process π will be transformed into the identity ι, thus the reverse of the sequence of adjacent swaps is a minimal decomposition of π.

$RandBS$ has been proved to have $\Theta(n^2)$ complexity. For further implementation details, proofs of correctness and complexity we refer the interested reader to [22].

3   Permutations with Repetition

3.1   Motivations and Preliminary Definitions

The search space of permutations arises in a variety of combinatorial problems such as, just to name a few: the permutation flowshop scheduling problem, the linear ordering problem, the quadratic assignment problem and the traveling salesman problem. Without loss of generality, an n-length permutation is an ordering of the set $\{1, \ldots, n\}$, thus the items in this ordering are all different from each other.

However, there exist other important combinatorial problems for which it is required that some items can appear several times in the ordering. For instance, in the job shop scheduling problem [11], the items are the jobs to be scheduled and repeated items accommodate the fact that some jobs need to be processed on more than one machine. Repeated items also allow to handle some partitioning problems, such as the balanced multiway graph partitioning problem [17].

We can encode solutions to these problems by means of *permutations with repetition*, i.e., orderings of a given multiset.

A multiset M is a collection of possibly repeated items, the size (or cardinality) of the collection is denoted by $|M|$, and its support $Supp(M)$ is the set of all different items appearing in M. For example, the multiset $M = \{1, 1, 2, 2, 3, 3\}$ has cardinality $|M| = 6$ and support $Supp(M) = \{1, 2, 3\}$.

Given a multiset M with support $\{1, \ldots, n\}$ and cardinality $q > n$, a permutation with repetition of M is an ordering of the q items in M. We denote by $\mathcal{R}_M$ the set of all the permutations with repetition of M. Considering the multiset M in the previous example, a possible permutation with repetition is $x = \langle 2, 1, 3, 3, 2, 1 \rangle$.

The search space $\mathcal{R}_M$ has size

$$|\mathcal{R}_M| = \frac{q!}{\prod_{i \in Supp(M)} m_M(i)!}, \tag{7}$$

where $m_M(i)$ is the multiplicity of the item i in M, i.e., the number of times i appears in M. Therefore, though $|\mathcal{R}_M| < |\mathcal{S}(q)|$, the size of the search space is anyway exponential with respect to the length of the orderings. This is the main reason of why the combinatorial problems with solutions in $\mathcal{R}_M$ are usually NP-hard.

For the sake of readability, in the rest of the paper, we will use the acronym *PwR* in place of the phrase "permutation with repetition".

3.2   Discrete Operators for Permutations with Repetition

Differently from classical permutations, it is not apparent how to define an internal operation on $\mathcal{R}_M$ which obeys to the group properties. As a consequence, it is not possible to directly use the discrete algebraic operators as defined in Sect. 2.

Anyway, the same simple search moves considered for permutations, i.e., swaps of adjacent items, can be used to move between permutations with repetition. Indeed, all the PwRs in $\mathcal{R}_M$ can be thought as vertices of a search space graph where, as before, its arcs are labelled by adjacent swap moves. Hence, the solutions $x, y \in \mathcal{R}_M$ are neighbors to each other if and only if y can be obtained from x (or vice versa) by swapping two adjacent items in x (or y).

By recalling that the adjacent swap move between the i-th and $(i + 1)$-th items (of a normal permutation, but also of a PwR) can be represented as the very simple permutation $\sigma_i \in ASW$ defined in Eq. (6), we have that a path between two given PwRs $x, y \in \mathcal{R}_M$ is a composition of adjacent swaps, i.e., a generic $|M|$-length permutation $\pi \in \mathcal{S}(|M|)$. Clearly, in this space we do not have the dichotomy observed in the Symmetric group $\mathcal{S}(n)$ that is: solutions and paths between solutions have different representations. Solutions are elements of $\mathcal{R}_M$, while paths/moves between solutions are elements of $\mathcal{S}(|M|)$.

The absence of the solution-move dichotomy does not allow to use the same algebraic definitions given in Sect. 2. Nevertheless, we can exploit the graph structure of $\mathcal{R}_M$ in order to derive reasonable definitions for the discrete sum, difference and scalar multiplication operators. These definitions are in line with the geometrical interpretations given in the previous section.

Discrete Sum. The discrete sum operator $\boxplus : \mathcal{R}_M \times \mathcal{S}(|M|) \to \mathcal{R}_M$ which, given a solution $x \in \mathcal{R}_M$ and a move $\pi \in \mathcal{S}(|M|)$, produces the new solution $y = x \boxplus \pi$ by applying to x all the adjacent swap moves appearing in a minimal decomposition of π in terms of ASW.

Discrete Difference. The discrete difference operator $\boxminus : \mathcal{R}_M \times \mathcal{R}_M \to \mathcal{S}(|M|)$ applied to two solutions $x, y \in \mathcal{R}_M$ produces the permutation $\pi = x \boxminus y$ whose minimal decomposition in terms of ASW is formed by the sequence of adjacent swaps that transform y into x. It is interesting to note that, similarly to what happens in the classical Euclidean space, $\boxplus$ and $\boxminus$ are consistent to each other, i.e., for any $x, y \in \mathcal{R}_M$, $x \boxplus (y \boxminus x) = x$.

Discrete Scalar Multiplication. Regarding the scalar multiplication, let observe that practically it is only used to scale-down a move or path in the space[1]. With this regard, see also the geometric interpretation of ADE in Sect. 2.2. Since a move in the search space of PwRs is a normal permutation, we can use unmodified the operator $\odot$ defined in Sect. 2 for the permutation space.

3.3   Implementation of the Discrete Operators

The definition previously given for $\boxplus$ actually indicates also its implementation. As noted in Sect. 2.3, decomposing the permutation costs $\Theta(|M|^2)$. Luckily, given $x \in \mathcal{R}_M$ and $\pi \in \mathcal{S}(|M|)$, it is possible to compute $x \boxplus \pi$ in linear time without decomposing π. Formally, by denoting with $x(i)$ the i-th item of the PwR x, we have that:

$$(x \boxplus \pi)(i) = x(\pi(i)). \tag{8}$$

It is easy to see that applying Eq. (8) to any item $i \in \{1, \ldots, |M|\}$ is equivalent to sequentially applying to x all the adjacent swaps in a decomposition of π.

Let also note that the operator $\boxplus$ is actually a (right) *group action* [18] of the Symmetric group $\mathcal{S}(|M|)$ on the set $\mathcal{R}_M$. Indeed, by using the Polish notation for the sake of readability, it is easy to verify that $\boxplus$ satisfies the two axioms of the (right) group action functions [18]: (i) $\boxplus(x, \iota) = x$ for all $x \in \mathcal{R}_M$, and (ii) $\boxplus(x, \pi \circ \sigma) = \boxplus(\boxplus(x, \pi), \sigma)$ for any $\pi, \sigma \in \mathcal{S}(|M|)$ and $x \in \mathcal{R}_M$.

For the discrete difference $\boxminus$, we first need to define the *canonical PwR* $e \in \mathcal{R}_M$ as the ordering of M whose items are increasingly sorted. For instance, given $M = \{1, 1, 2, 2, 3, 3\}$, its canonical PwR is $e = \langle 1, 1, 2, 2, 3, 3 \rangle$.

Furthermore, let observe that the concept of inversion, introduced in Sect. 2.3, is also defined on the permutations with repetition, and e is the only PwR without inversions.

Therefore, it is possible to use *RandBS* – or, if randomness is not required, any other bubble-sort variant – to sort any PwR x, towards the canonical PwR e, by using an optimal number of adjacent swaps. The optimality derives from the facts that: (i) bubble-sort schemes are known to be optimal when all items are different, and (ii) useless adjacent swaps between equivalent items in a PwR are avoided because the pairs of equivalent items cannot form inversions.

Hence, we are now able to find the sequence of adjacent swaps for moving from any PwR x towards e, i.e., we are able to compute $e \boxminus x$. Moreover, by

---

[1] Even in the Euclidean space $\mathbb{R}^n$, multiplying a vector by a scalar has a geometric meaning only if we interpret this vector as a proper free vector and not as a point in the space.

observing the commutative diagram depicted in Fig. 1, we can generalize the computation $x \boxminus y$ to any $x, y \in \mathcal{R}_M$. In this diagram, any arrow connects two PwRs and is labelled with the permutation which encodes the sequence of adjacent swaps that transform the tail of the arrow into its head. Equivalently, the label of the arrow is the difference between the head PwR and the tail PwR.

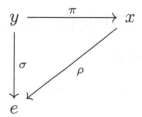

Fig. 1. Commutative diagram showing how to compute $\pi = x \boxminus y$

Since we know how to compute $\sigma = e \boxminus y$ and $\rho = e \boxminus x$, we can now define the difference between two generic PwRs as:

$$x \boxminus y = \pi = \sigma \circ \rho^{-1}. \tag{9}$$

4   Algebraic Differential Evolution for Permutations with Repetition

In this section we define an Algebraic Differential Evolution scheme for Permutations with Repetition (ADE-PR) which is based on the ADE scheme described in Sect. 2.2 and the discrete operators for the PwR representation depicted in Sect. 3.

ADE-PR evolves a population of N permutations with repetition by means of the genetic operators: differential mutation, crossover and selection. Its working scheme, depicted in Algorithm 1, is similar to those of ADE and classical DE. The main difference is that the population of ADE-PR is composed by individuals represented as permutations with repetition.

ADE-PR optimizes a given objective function f defined on the search space $\mathcal{R}_M$. Its control parameters are: the population size N, the scale factor $F \in [0, 1]$ and the crossover strength $CR \in [0, 1]$ (the latter may not be present depending on the chosen crossover operator).

In Algorithm 1, the population is randomly initialized in lines 2–3, then the evolution is performed in the main cycle in lines 4–12 until a given termination criterion is satisfied. For each population individual x_i, a mutant v_i is generated in line 6 by exploiting the differential mutation scheme which is implemented by means of the discrete operators for PwRs previously introduced. Then, a trial PwR u_i is obtained, in line 7, by hybridizing x_i and v_i by means of a chosen

---

Algorithm 1. Main scheme of ADE-PR

1: **function** ADE-PR($f : \mathcal{R}_M \to \mathbb{R}, N \in \mathbb{N}^+, F \in [0,1], CR \in [0,1]$)
2:　　**for** $i \leftarrow 1, \ldots, N$ **do**
3:　　　　$x_i \leftarrow$ randomly sample a PwR from $\mathcal{R}_M$
4:　　**while** termination criterion is not satisfied **do**
5:　　　　**for** $i \leftarrow 1, \ldots, N$ **do**
6:　　　　　　$v_i \leftarrow x_{r_0} \boxplus F \odot (x_{r_1} \boxminus x_{r_2})$
7:　　　　　　$u_i \leftarrow \mathtt{Crossover}(x_i, v_i, CR)$
8:　　　　　　Optionally apply a local search procedure on u_i
9:　　　　**for** $i \leftarrow 1, \ldots, N$ **do**
10:　　　　　　**if** $f(u_i) < f(x_i)$ **then**
11:　　　　　　　　$x_i \leftarrow u_i$
12:　　　　Optionally perform a population restart
13:　　**return** x_{best}
14: **end function**

---

crossover operator working on the PwR representation (like, for instance, GOX, GPMX and PPX [8,9]). In lines 9–10, if u_i is fitter than x_i, it enters the next-generation population by replacing x_i.

Moreover, it is possible to integrate into the ADE-PR scheme: (i) a local search scheme, purposely defined for the problem at hand, in order to refine the search, and (ii) a restart procedure which is often useful in combinatorial problems whose search space is finite.

Note also that the parameters F and CR can be self-adapted during the evolution using one of the many self-adaptive DE schemes in the literature.

The key operator of ADE-PR is the newly introduced differential mutation scheme which directly works with permutations with repetition. The expression in line 6 of Algorithm 1 can be interpreted as follows. The mutant v_i is generated by applying to the PwR x_{r_0} a sequence of adjacent swap moves which is a prefix of the sequence of moves that transform the PwR x_{r_2} into x_{r_1}. Clearly, the length of the prefix is regulated by the scale factor $F \in [0,1]$. This interpretation is in line with what happens in the continuous DE and for ADE in a search space representable as a finitely generated group.

5　ADE-PR for the Job Shop Scheduling Problem

As a case study, we describe an implementation of ADE-PR for solving the Job Shop Scheduling Problem (JSSP). The resulting algorithm, called ADE-PR-JSSP, follows the scheme depicted in Sect. 4, i.e., it starts with a population of randomly initialized PwRs which are evolved by means of the following operators: discrete differential mutation, GOX crossover [9], selection, local search for the JSSP, and restart procedure. Furthermore, the parameter F used in the discrete differential mutation is self-adapted by means of the jDE method [10], while the GOX crossover, as defined in [9], has no parameter.

In the next subsections we describe: the definition of the JSSP, the procedure for converting a permutation with repetition to a feasible JSSP schedule, the local search operator and the restart scheme adopted in ADE-PR-JSSP.

5.1   Definition of the Problem

The Job Shop Scheduling Problem (JSSP) is an important scheduling problem with many applications in the manufacturing and service industry [12,13].

An instance of the JSSP is defined in terms of a set $\mathcal{J}$ of n jobs $J_1, \ldots, J_n$ and a set $\mathcal{M}$ of m machines $\mu_1, \ldots, \mu_m$. Each job J_i is composed by m operations $O_{i1}, \ldots, O_{im}$. Every operation O_{ij} has a processing time p_{ij} and has to be executed by the machine $\mu_{ij} \in \mathcal{M}$. All the operations within a given job are linearly ordered, while no constraint is defined among operations belonging to different jobs. The set of all the operations is denoted by $\mathcal{O}$.

A feasible schedule s consists in assigning to each operation $O_{ij} \in \mathcal{O}$ a start time s_{ij} such that the following constraints are satisfied: for each $i = 1, \ldots, n$ and $j = 1, \ldots, m-1$,

$$s_{ij} \leq s_{i,j+1}, \tag{10}$$

and, for each $O_{ij}, O_{hk} \in \mathcal{O}$ with $\mu_{ij} = \mu_{hk}$,

$$s_{ij} \geq e_{hk} \text{ or } s_{hk} \geq e_{ij}, \tag{11}$$

where $e_{ij} = s_{ij} + p_{ij}$ is the end time of the operation O_{ij}.

A feasible schedule s is optimal if it optimizes a given objective function. In this paper, the aim is to minimize the makespan

$$C_{max}(s) = \max_{ij} e_{ij}. \tag{12}$$

The JSSP has been approached using a variety of different techniques. In the recent survey [11] many evolutionary and meta-heuristic approaches to solve the JSSP are described: Particle Swarm Optimization, Ant Colony Optimization, Variable Neighborhood Search, Tabu Search, Genetic Algorithms, and several others.

5.2   From a Permutation with Repetition to a JSSP Schedule

The solutions of ADE-PR-JSSP are represented as PwRs over the multiset $M_{m,n}$, whose support is $\{1, \ldots, n\}$, and such that every item in $M_{m,n}$ has multiplicity m. Hence, ADE-PR-JSSP navigates the search space of the permutations with repetition in $\mathcal{R}_{M_{m,n}}$.

This representation, called *operation-based representation* [12], was firstly introduced by [8] and has the important property that it generates only feasible solutions.

The operation-based representation is based on the fact that each operation $O_{ij} \in \mathcal{O}$ can be uniquely identified by the integer number $(i-1)m+j$. Therefore, a PwR $x \in \mathcal{R}_{M_{m,n}}$ is decoded to a JSSP schedule by using a two-phase procedure.

In the first phase, a permutation $\pi_x \in \mathcal{S}(mn)$, representing a total order $\prec_x$ among the operations in $\mathcal{O}$, is built from x as follows.

For each $h = 1, \ldots, mn$, let $j = x(h)$ be the h–th item of x and let k be the number of items x_l, for $1 \leq l \leq h$, such that $x_l = j$, then $\pi_x(h)$ is set to $(j-1)m + k$. Then, $\pi_x(h)$ corresponds to the operation $O_{j,k}$.

It is easy to see that $\prec_x$ respects the constraint (10) by construction. Indeed $O_{ij} \prec_x O_{i,j+1}$, for each pair of indices i and $j < m$. Moreover, for each pair of operations $O_{ij}, O_{hk} \in \mathcal{O}$ assigned to the same machine, $\prec_x$ states in which order the two operations have to be executed.

The second phase assigns to each operation O_{ij} a start time s_{ij} as the maximum among the end time of $O_{i,j-1}$ and the end times of all the operations O_{hk} preceding O_{ij}, with respect to $\prec_x$, and such that $\mu_{ij} = \mu_{hk}$.

The obtained schedule is feasible and respects the precedence relations induced by π_x. Therefore, by calling the conversion procedure as *GenerateSchedule* and given a PwR $x \in \mathcal{R}_{M_{m,n}}$, we have that the fitness of x in ADE-PR-JSSP is the makespan of the corresponding schedule, i.e., C_{max} (*GenerateSchedule*(x)).

5.3   Local Search for the JSSP

We have designed ADE-PR-JSSP in such a way that every trial individual $u_i \in \mathcal{R}_{M_{m,n}}$, at every generation of the algorithm, undergoes a local search procedure with probability p_{LS}.

Before applying the local search, the trial individual, represented as a PwR, is first converted to a schedule by means of the procedure described in Sect. 5.2.

The local search is based on the neighborhood $\mathcal{N}^*$, as described in [21], and works as follows. At each iteration the *critical path* of the current schedule s is computed. This path is the sequence of consecutive operations (where the end time of any operation coincides with the start time of the successive operation in the path) which has the maximum completion time (which corresponds to the makespan of s). Then, the blocks of consecutive operations assigned to the same machine are detected in the critical path. For each block B, two swaps are tried: one exchanges the first two operations in B, while the other exchanges the last two operations. The swap which most reduces the makespan is performed and the schedule is updated accordingly. If no swap produces a better makespan, the local search terminates.

Now, the local optimal schedule is converted back to a PwR and replaces the seed individual u_i in the population of ADE-PR-JSSP. The conversion can be easily implemented by considering a topological sorting in the precedence graph of the local optimal schedule.

After some preliminary experiments, we set the probability p_{LS} to apply the local search as

$$p_{LS}(t) = \frac{t}{T} \cdot p_{LS}^{end} + \left(1 - \frac{t}{T}\right) \cdot p_{LS}^{start}, \tag{13}$$

where t is the current computational time, T is the budget for the execution time, p_{LS}^{start} is the probability of applying the local search at time $t = 0$, and $p_{LS}^{end} >$

p_{LS}^{start} is the probability at time $t = T$. Hence, the local search is progressively applied more often as time passes. This behavior should favor exploration in the earlier phase of the evolution, while exploitation is intensified with the passing of time.

5.4 Restart Scheme

The restart mechanism is implemented by replacing all the population individuals, except the best one, with new randomly generated PwRs.

A restart is performed when the algorithm has not been able to improve its best solution so far after $T \cdot r_{restart}$ seconds, where T is the total allotted running time and $r_{restart} < 1$ is the parameter which regulates how often this operation should be performed at most.

6 Experiments

ADE-PR-JSSP has been experimentally validated on some commonly adopted benchmarks for the JSSP, namely: the ft, la, and orb benchmark suites [15][2]. The benchmarks contain a total of 53 JSSP instances with $n \cdot m$ ranging from 36 to 300.

After some preliminary experiments, the population size has been set to $N = 25$ individuals, the range for the application probability of the local search has been set using $p_{LS}^{start} = 0$ and $p_{LS}^{end} = 1$, while the restart parameter $r_{restart}$ has been set to 0.1.

The executions of ADE-PR-JSS have been carried out on a machine equipped with the Intel Xeon CPU E5-2620 v4 clocking at 2.10 GHz. Every execution terminates after a time budget of $T = 4mn$ seconds has been exhausted. Moreover, $R = 15$ executions per instance have been run.

The presentation of the experimental results is divided in three groups, according to the values of mn: Table 1 refers to the instances with $nm < 100$, Table 2 to those with $nm = 100$, and Table 3 to the remaining instances. In these three tables we present, for each instance: the sizes n and m, the average (Avg_i) and minimum (Min_i) fitness values obtained by ADE-PR-JSSP in the R runs, the known optimal value (Opt_i) for the instance (taken from the recently published survey paper [15]), and the average relative percentage deviation computed as $ARPD_i = 100 \times \frac{Avg_i - Opt_i}{Opt_i}$. Moreover, the minimum Min_i is reported in boldface when it matches the known optimal value Opt_i.

Interestingly, in 38 out of 53 instances, ADE-PR-JSSP reached the optimal value at least once, while, in 22 instances, this happened in all the executions.

In particular, the results provided in Table 1 refer to small JSSP instances, where ADE-PR-JSSP has been always able to find the optimal value, and for 7 of such instances this happened in all the executions. Indeed, the average ARPD for this set of instances is rather small, i.e., 0.167%.

---

[2] These JSSP instances can be downloaded from the website http://jobshop.jjvh.nl.

Table 1. Experimental results on instances with $nm < 100$

Instance	n	m	Avg	ARPD	Min	Opt
ft06	6	6	55.000	0.000	**55**	55
la01	10	5	666.000	0.000	**666**	666
la02	10	5	660.133	0.784	**655**	655
la03	10	5	602.867	0.983	**597**	597
la04	10	5	590.400	0.068	**590**	590
la05	10	5	593.000	0.000	**593**	593
la06	15	5	926.000	0.000	**926**	926
la07	15	5	890.000	0.000	**890**	890
la08	15	5	863.000	0.000	**863**	863
la09	15	5	951.000	0.000	**951**	951
la10	15	5	958.000	0.000	**958**	958

Table 2. Experimental results on instances with $nm = 100$

Instance	n	m	Avg	ARPD	Min	Opt
ft10	10	10	937.933	0.853	**930**	930
ft20	20	5	1174.600	0.824	**1165**	1165
la11	20	5	1222.000	0.000	**1222**	1222
la12	20	5	1039.000	0.000	**1039**	1039
la13	20	5	1150.000	0.000	**1150**	1150
la14	20	5	1292.000	0.000	**1292**	1292
la15	20	5	1207.000	0.000	**1207**	1207
la16	10	10	949.800	0.508	**945**	945
la17	10	10	785.267	0.162	**784**	784
la18	10	10	848.667	0.079	**848**	848
la19	10	10	845.667	0.435	**842**	842
la20	10	10	906.667	0.517	**902**	902
orb01	10	10	1078.267	1.819	1064	1059
orb02	10	10	890.800	0.315	889	888
orb03	10	10	1019.000	1.393	**1005**	1005
orb04	10	10	1018.267	1.320	1011	1005
orb05	10	10	891.400	0.496	889	887
orb06	10	10	1025.800	1.564	1021	1010
orb07	10	10	401.667	1.175	**397**	397
orb08	10	10	906.800	0.868	**899**	899
orb09	10	10	943.467	1.014	**934**	934
orb10	10	10	944.000	0.000	**944**	944

Table 3. Experimental results on instances with $nm > 100$

Instance	n	m	Avg	ARPD	Min	Opt
la21	15	10	1059.933	1.332	1047	1046
la22	15	10	930.133	0.338	**927**	927
la23	15	10	1032.000	0.000	**1032**	1032
la24	15	10	941.667	0.713	938	935
la25	15	10	982.867	0.600	982	977
la26	20	10	1218.000	0.000	**1218**	1218
la27	20	10	1257.533	1.825	1242	1235
la28	20	10	1221.533	0.455	**1216**	1216
la29	20	10	1190.067	3.304	1174	1152
la30	20	10	1355.000	0.000	**1355**	1355
la31	30	10	1784.000	0.000	**1784**	1784
la32	30	10	1850.000	0.000	**1850**	1850
la33	30	10	1719.000	0.000	**1719**	1719
la34	30	10	1721.000	0.000	**1721**	1721
la35	30	10	1888.000	0.000	**1888**	1888
la36	15	15	1286.267	1.441	1278	1268
la37	15	15	1435.600	2.763	1418	1397
la38	15	15	1210.867	1.243	1202	1196
la39	15	15	1249.400	1.330	1246	1233
la40	15	15	1240.133	1.484	1228	1222

Table 2 shows that on the 22 selected instances with $nm = 100$ the algorithm has been able to find the optimal value: at least once on 17 instances, and in all the executions in 6 cases. As expected, the average ARPD for this second set of instances, 0.606%, is larger than the previous, but anyway close to 0.

In Table 3 it is possible to see that ADE-PR-JSSP reached the known optimal value: at least once in half the instances (10 out of 20), and in all the executions for 8 of them. Therefore, the average ARPD for this last set of instances is slightly larger than the other: 0.841%. Moreover, Table 3 also shows that the instances with $m = 15$ are much harder and the average ARPD restricted to this subset raises to 1.652%.

Summarizing, the overall ARPD obtained by averaging on all the 53 instances is 0.604, thus promoting the proposed approach as a method competitive with respect to the known values for the considered benchmarks.

7   Conclusion and Future Work

In this paper, we have extended the algebraic framework for evolutionary computation previously proposed in [22] in order to handle the search space of permutations with repetition. The newly proposed discrete operators allowed to design an Algebraic Differential Evolution called ADE-PR which can be applied to any combinatorial optimization problem whose solutions may be represented as permutations of possibly repeated items.

In particular, ADE-PR has been devised for the Job Shop Scheduling Problem (JSSP). In order to validate the effectiveness of the proposed approach, experiments have been on a set of widely adopted benchmark instances for the JSSP. The experimental results show that our proposal is competitive with respect to the known optimal objective values for the considered benchmarks.

Possible future lines of research are: apply ADE-PR to partitioning problems; use simple search moves other than the swaps of adjacent items; design other algebraic evolutionary algorithms, like the APSO [23], in order to work with permutations with repetition; and generalize the approach to other search spaces, by means of the algebraic concept of *group action*, in order to see the deployed discrete operators as projections from a known space which can be represented as a group to other more general combinatorial spaces.

Acknowledgement. The research described in this work has been partially supported by: "Università per Stranieri di Perugia – Finanziamento per Progetti di Ricerca di Ateneo –PRA 2020", and by RCB-2015 Project "Algoritmi Randomizzati per l'Ottimizzazione e la Navigazione di Reti Semantiche" and RCB-2015 Project "Algoritmi evolutivi per problemi di ottimizzazione combinatorica" of Department of Mathematics and Computer Science of University of Perugia.

References

1. Baioletti, M., Milani, A., Santucci, V.: Algebraic crossover operators for permutations. In: 2018 IEEE Congress on Evolutionary Computation (CEC 2018), pp. 1–8 (2018). https://doi.org/10.1109/CEC.2018.8477867
2. Baioletti, M., Milani, A., Santucci, V.: A new precedence-based ant colony optimization for permutation problems. In: Shi, Y., et al. (eds.) SEAL 2017. LNCS, vol. 10593, pp. 960–971. Springer, Cham (2017). https://doi.org/10.1007/978-3-319-68759-9_79
3. Baioletti, M., Milani, A., Santucci, V.: Automatic algebraic evolutionary algorithms. In: Pelillo, M., Poli, I., Roli, A., Serra, R., Slanzi, D., Villani, M. (eds.) WIVACE 2017. CCIS, vol. 830, pp. 271–283. Springer, Cham (2018). https://doi.org/10.1007/978-3-319-78658-2_20
4. Baioletti, M., Milani, A., Santucci, V.: Learning Bayesian networks with algebraic differential evolution. In: Auger, A., Fonseca, C.M., Lourenço, N., Machado, P., Paquete, L., Whitley, D. (eds.) PPSN 2018. LNCS, vol. 11102, pp. 436–448. Springer, Cham (2018). https://doi.org/10.1007/978-3-319-99259-4_35
5. Baioletti, M., Milani, A., Santucci, V.: MOEA/DEP: an algebraic decomposition-based evolutionary algorithm for the multiobjective permutation flowshop scheduling problem. In: Liefooghe, A., López-Ibáñez, M. (eds.) EvoCOP 2018. LNCS, vol. 10782, pp. 132–145. Springer, Cham (2018). https://doi.org/10.1007/978-3-319-77449-7_9

6. Baioletti, M., Milani, A., Santucci, V.: Variable neighborhood algebraic differential evolution: an application to the linear ordering problem with cumulative costs. Inf. Sci. **507**, 37–52 (2020). https://doi.org/10.1016/j.ins.2019.08.016
7. Baioletti, M., Milani, A., Santucci, V., Bartoccini, U.: An experimental comparison of algebraic differential evolution using different generating sets. In: Proceedings of the Genetic and Evolutionary Computation Conference Companion, GECCO 2019, pp. 1527–534 (2019). https://doi.org/10.1145/3319619.3326854
8. Bierwirth, C.: A generalized permutation approach to job shop scheduling with genetic algorithms. Oper. Res. Spektrum **17**(2), 87–92 (1995)
9. Bierwirth, C., Mattfeld, D.C., Kopfer, H.: On permutation representations for scheduling problems. In: Voigt, H.-M., Ebeling, W., Rechenberg, I., Schwefel, H.-P. (eds.) PPSN 1996. LNCS, vol. 1141, pp. 310–318. Springer, Heidelberg (1996). https://doi.org/10.1007/3-540-61723-X_995
10. Brest, J., Greiner, S., Boskovic, B., Mernik, M., Zumer, V.: Self-adapting control parameters in differential evolution: a comparative study on numerical benchmark problems. IEEE Trans. Evol. Comput. **10**(6), 646–657 (2006)
11. Çalış, B., Bulkan, S.: A research survey: review of AI solution strategies of job shop scheduling problem. J. Intell. Manuf. **26**(5), 961–973 (2015)
12. Cheng, R., Gen, M., Tsujimura, Y.: A tutorial survey of job-shop scheduling problems using genetic algorithms, Part I: representation. Comput. Ind. Eng. **30**(4), 983–997 (1996)
13. Cheng, R., Gen, M., Tsujimura, Y.: A tutorial survey of job-shop scheduling problems using genetic algorithms, Part II: hybrid genetic search strategies. Comput. Ind. Eng. **36**(2), 343–364 (1999)
14. González, M.Á., Oddi, A., Rasconi, R.: Multi-objective optimization in a job shop with energy costs through hybrid evolutionary techniques. In: Proceedings of the 27th International Conference on Automated Planning and Scheduling (2017)
15. van Hoorn, J.J.: The current state of bounds on benchmark instances of the job-shop scheduling problem. J. Sched. **21**(1), 127–128 (2018)
16. Jovanovski, J., Arsov, N., Stevanoska, E., Siljanoska Simons, M., Velinov, G.: A meta-heuristic approach for RLE compression in a column store table. Soft Comput. **23**(12), 4255–4276 (2019)
17. Kim, J., Hwang, I., Kim, Y.H., Moon, B.R.: Genetic approaches for graph partitioning: a survey. In: Proceedings of the GECCO 2011, pp. 473–480. ACM (2011)
18. Lang, S.: Algebra, vol. 211. Springer, Heidelberg (2002). https://doi.org/10.1007/978-1-4613-0041-0
19. Milani, A., Santucci, V.: Asynchronous differential evolution. In: Proceedings of 2010 IEEE Congress on Evolutionary Computation (CEC 2010), pp. 1–7 (2010). https://doi.org/10.1109/CEC.2010.5586107
20. Moraglio, A., Kim, Y.H., Yoon, Y., Moon, B.R.: Geometric crossovers for multiway graph partitioning. Evol. Comput. **15**(4), 445–474 (2007)
21. Nowicki, E., Smutnicki, C.: An advanced tabu search algorithm for the job shop problem. J. Sched. **8**(2), 145–159 (2005)
22. Santucci, V., Baioletti, M., Milani, A.: Algebraic differential evolution algorithm for the permutation flowshop scheduling problem with total flowtime criterion. IEEE Trans. Evol. Comput. **20**(5), 682–694 (2016). https://doi.org/10.1109/TEVC.2015.2507785
23. Santucci, V., Baioletti, M., Milani, A.: Tackling permutation-based optimization problems with an algebraic particle swarm optimization algorithm. Fundamenta Informaticae **167**(1–2), 133–158 (2019). https://doi.org/10.3233/FI-2019-1812

24. Santucci, V., Baioletti, M., Di Bari, G., Milani, A.: A binary algebraic differential evolution for the multidimensional two-way number partitioning problem. In: Liefooghe, A., Paquete, L. (eds.) EvoCOP 2019. LNCS, vol. 11452, pp. 17–32. Springer, Cham (2019). https://doi.org/10.1007/978-3-030-16711-0_2
25. Santucci, V., Ceberio, J.: Using pairwise precedences for solving the linear ordering problem. Appl. Soft Comput. **87** (2020). https://doi.org/10.1016/j.asoc.2019.105998
26. Santucci, V., Milani, A.: Particle swarm optimization in the EDAs framework. In: Gaspar-Cunha, A., Takahashi, R., Schaefer, G., Costa, L. (eds.) Soft Computing in Industrial Applications, pp. 87–96. Springer, Heidelberg (2011). https://doi.org/10.1007/978-3-642-20505-7_7
27. Sörensen, K.: Distance measures based on the edit distance for permutation-type representations. J. Heuristics **13**(1), 35–47 (2007)

A Comparison of Genetic Representations for Multi-objective Shortest Path Problems on Multigraphs

Lilla Beke[✉], Michal Weiszer, and Jun Chen

School of Engineering and Materials Science, Queen Mary University of London,
Mile End Road, London, UK
l.beke@qmul.ac.uk

Abstract. The use of multi-graphs in modelling multi-objective transportation problems is gaining popularity, necessitating the consideration of the Multi-objective Shortest Path Problem (MSPP) on multigraphs. This problem is encountered in time-dependent vehicle routing, multimodal transportation planning and in optimising airport operations. This problem is more complex than the NP-hard simple graph MSPP, and thus approximate solution methods are needed to find a good representation of the true Pareto front in a given time budget. Evolutionary algorithms have been applied with success to the simple graph MSPP, however their performances on multigraph MSPP were not systematically investigated. To this aim, we extend the most popular genetic representations to the multigraph case and compare the achieved performances. We find that the priority based encodings outperform the direct ones with purely random initialisation. We further introduce a novel heuristic initialisation technique, that is generic enough for many representations, and that further improves the convergence speed and solution quality of the algorithms. The results are encouraging for later application to the time constrained multigraph MSPP.

Keywords: Multi-objective shortest path problems · Multigraphs · Genetic representation techniques · Heuristic initialisation

1 Introduction

There is substantial evidence [6,9,15,25] that representing routing problems as multigraphs offers benefits with regards to time, cost, environmental impact and flexibility in multiple practical settings. The availability of multiple parallel edges between pairs of nodes is rooted in the multi-objective nature of the problems. If there are multiple ways to traverse a section of the route offering different trade-offs between the objectives, they should all be modelled in the optimisation

This work is supported in part by the Engineering and Physical Sciences Research Council.

L. Paquete and C. Zarges (Eds.): EvoCOP 2020, LNCS 12102, pp. 35–50, 2020.
https://doi.org/10.1007/978-3-030-43680-3_3

process in order to find the set of Pareto-optimal solutions. The parallel edges in practice might correspond to different physical routes as in multi-objective vehicle routing problems [9,11,25], different modes of transport [15], or the same physical route traversed with different speed profiles such as in the airport ground movement problem [6].

The multi-objective shortest path problem (MSPP) on multigraphs can be reduced to two NP-hard problems, the MSPP on simple graphs and the Fixed Sequence Arc Selection Problem (FSASP) [11]. In the FSASP the sequence of nodes to be traversed is fixed but there are multiple alternative routes (parallel edges) between any neighbouring nodes in the sequence. For the simple graph MSPP the number of solutions grows exponentially with the number of nodes in the worst case, and for the multigraph MSPP getting close to this worst case is even more probable. Consequently, the running times of exact solution approaches for the multigraph MSPP are often unacceptable in real applications.

Metaheuristics and in particular genetic algorithms have been used with success [7,18,19] to solve simple graph shortest path problems in the context of communication and transportation. We extend four of the main genetic representation methods for the multigraph case. These are the direct variable length [4], direct fixed length [16], random key [12] and integer valued priority [19] representations. A comparison of the representations is an important step towards the design of an efficient metaheuristic algorithm for the multigraph MSPP.

Heuristic initialisation techniques have been used with success in different combinatorial optimisation problems, as they can lead to quicker convergence by starting the evolutionary process with an initial population of higher quality. However, this approach is mostly overlooked when solving shortest path problems, the main concern being premature convergence caused by a lack of sufficient diversity. Premature convergence can be avoided by introducing enough randomness into the population, while still preserving higher quality compared to a purely random population. We introduce a new heuristic initialisation method applicable for all four mentioned representations to speed up convergence.

The problem is described in Sect. 2. Related work on representation methods for the simple and multigraph MSPP, and initialisation techniques are summarised in Sect. 3. Section 4 describes our approach to extend the main representations to the multigraph case and the heuristic initialisation technique. In Sect. 5, the numerical experiments and their results are presented. Conclusion is drawn and future research directions are described in Sect. 6.

2   Problem Description

The multigraph MSPP is defined by a multigraph network $G = (V, E)$ and a multi-dimensional cost-vector associated with each edge in G. The network is assumed to be undirected in this paper. $V = 1, ..., n$ represents the set of nodes, and E the set of edges. Given that G is a multigraph, there might be multiple edges in E connecting the same nodes. For this reason an edge in the network is denoted by $e = (v, u, i)$, where $u, v \in V$ and i is a parallel edge index, that differentiates between the edges with the same endpoints u and v. These edges are numbered starting from 1 to the number of parallel edges between the two nodes $l(u, v)$. There are two special nodes $O, D \in V$, the origin and destination nodes respectively. The costs associated with each edge according to k objectives considered are given by a k-dimensional cost-vector $\overline{cost(e)} = (c_1(e), c_2(e), ..., c_k(e))$. For a valid path P between O and D, the corresponding cost-vector can be calculated according to Eq. (1).

$$C(P) = \sum_{e \in P} \overline{cost(e)} \qquad (1)$$

We are looking for the minimum cost path between O and D considering all k objectives. The solution is the set of valid paths with non-dominated cost-vectors, i.e. the Pareto optimal solutions. A solution path P_1 is said to be non-dominated if another solution path that is at least as good as P_1 according to all k objectives and better according to at least one objective does not exist.

3 Related Work

3.1 Genetic Representations for Paths in Simple Graphs

Multiple genetic representation methods have been proposed for the simple graph version of the problem, the most popular ones can be classified as direct representations priority-based representations.

Direct Variable Length Encoding. In the simple graph problem candidate solutions need to encode a sequence of nodes in a graph. The most straightforward way of doing this is the direct variable length representation proposed in [24] for the single objective shortest path problem (SSPP). Chromosomes consist of lists of node IDs, that form a path starting with the origin node and ending with the destination node. An arbitrary list of nodes usually won't correspond to a feasible path in the graph, and this necessitates the use of problem specific genetic operators. The overlap-based crossover and random walk based mutation introduced by Ahn and Ramakrishna [4] are the most popular choice, and these were adapted by multiple authors for the multi-objective problem [7,17,18].

The main advantage of this representation is that it gives a one-to-one mapping. However, the operators might lead to loop formation, and thus offspring need to be checked and repaired after mutation and crossover. Also, according to [20,23] this representation is not suitable for large networks.

Direct Fixed Length Encoding. Another node ID based representation was proposed by Inagaki et al. [16]. The length of the chromosome equals n, the number of nodes in the network, and the node IDs are the numbers from 1 to n. The chromosomes incorporate a pointer to a neighbouring node for each node. This way the path is decoded by following the pointers until the destination is reached or a loop is formed. The locus of a gene corresponds to a node in the network with the same ID, and the value of the gene is the ID of a neighbour that is the next node in the path. The crossover operator applied in [16] is essentially a uniform crossover, which is deemed inconsistent and requires large population sizes [4, 20].

Integer Valued Priority Based Encoding. Gen et al. proposed a priority based encoding technique with integer priority values [13], which encodes the solution path indirectly through some guiding information.

Priority-based encoding is a permutation encoding, where a chromosome contains priority values for each node in the network. In this case, the priorities are integers from 1 to n. A path is decoded from the chromosome by starting at the origin node and appending to the path the node with the highest priority among the neighbours of the current node, given it is not yet in the path. If it is already in the path, the neighbour with the next highest priority is chosen instead.

The main advantage of this representation is that common crossover operators can be used and expected to work well, unlike in case of the direct representation. This is mainly because a random permutation of the priorities will always be decoded to some valid path starting from the origin node. In direct representation a random permutation of node IDs will most likely not correspond to a valid path, in the sense that the consecutive node IDs will not be adjacent in the network. This representation has been shown to perform well in comparison to the variable length direct representation for the bi-objective problem with Weight Mapping Crossover (WMX) proposed specifically for this problem [19]. WMX is an extension of the one-cut point crossover for permutation representation.

Random Key Based Encoding. Gen and Lin proposed another similar encoding technique by using floating-point numbers instead of integers as priorities [14], resulting in a random key representation for the single objective shortest path problem. Random keys were found to be a powerful method for permutation representation in other combinatorial optimisation problems [5]. The advantage of random key encoding is that even simpler operators might be used. In [14] arithmetical crossover is employed, where the offspring are calculated as the weighted average of the two parent vectors.

3.2   Genetic Representations for Paths in Multigraphs

For multigraph, not only the sequence of nodes needs to be specified, but also the index of a parallel edge between any two consecutive nodes. A multigraph network and a solution path within it is illustrated in Fig. 1. Although this is a relevant issue for a wider range of real-world problems, the only previous attempts to solve it to our knowledge are from the context of multi-modal transportation [3, 27].

Abbaspour and Samadzadegan extended the direct variable length representation to the multi-modal problem [3]. The length of the chromosomes are double

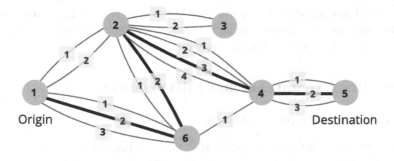

Fig. 1. An example of a multigraph network and a solution path from node 1 to node 5 in it, indicated by a wider line.

the number of nodes in the network. The genes at odd loci correspond to node IDs, just like in [4]. The genes at even loci are used to indicate the mode of travel to be used between the node IDs indicated by the neighbouring genes. The same crossover and mutation operators are used on the odd genes as in simple graph problem, and the even genes (modes of transport) only change with the odd ones.

Yu and Lu [27] propose a slightly different approach, where they only indicate the changes in the mode of travel and do not specify it for each edge separately. Genes indicating the mode of travel for the consecutive node IDs have a negative value, to differentiate them from the node IDs. However, only a limited number of parallel edges were used. When the number of parallel edges groves, the possible advantage of a shorter chromosome is diminished by the burden of maintaining the chromosome structure with more complicated operators. For this reason we propose a strategy that encodes the indices of parallel edges for each consecutive node in the paths, not just the changes.

3.3   Initialisation

Most mentioned works applied purely random initialisation. An exception is [17], where spatial information is utilised to guide the random walk further from the origin node and closer to the destination node. Spatial information however is not always available and in some cases might be misleading. Another example for heuristic initialisation can be found in [18], where shortest paths are found according to some weighted aggregation of smaller subsets of the objectives with Dijkstra's algorithm, and these are included in the initial population along with randomly generated chromosomes. The disadvantage of this approach is that it cannot be used with the priority-based representations.

A different heuristic initialisation approach was proposed for Particle Swarm Optimisation in [23] for solving the SSPP with a priority based representation method. They achieved decreased risk of loop formation by using the values of node IDs to reject backwards moves when decoding the solution path. Priorities were randomly assigned and node IDs were assigned in a way that the higher the

IDs are, the closer a node is to the destination. They rejected backwards moves in the decoding process, defined as moves that would lead to some given level of decrease in node IDs, when other moves were also available.

4   New Representations and Initialisation

4.1   Extension of Representations

Representations for the multigraph MSPP need to encode the parallel indices for each pair of consecutive nodes in a solution path, not just the node sequence. This is easier for direct representation than for the priority based ones if the network might contain varied numbers of parallel edges between pairs of nodes. The previously proposed approaches for encoding mode of transport in the multimodal transportation literature employ direct representation, which allows for including the parallel edge index directly. Our proposed way of representing parallel edges works with priority based representations that were previously not extended to the multigraph problem to our knowledge.

Using the multimodal transportation problem for illustrative purposes, the difficulty of applying the same approach for priority-based representation as in [3] is the following. Generally, there are different modes of transport (different number of parallel edges) available between pairs of nodes in the network. This means that if we assign a mode of transport to a pair of adjacent nodes randomly, it might not be feasible. In direct representation, such as [3], once the modes of transport are initialised to be feasible, the crossover and mutation operators do not ruin feasibility. One might try to follow a similar strategy with priority based representation and assign modes of transport to node IDs to be used when leaving the node. Then it might happen that a crossover results in a solution path in which a node u is followed by another node v that does not even appear in the paths encoded by the parents of the given chromosome. Additionally, the modes used for leaving u in the parents might not be available between u and v. This could be resolved with a repair mechanism, but here we propose an alternative method for representing the parallel indices that does not require repairing.

Instead of encoding the indices of parallel edges directly, we use a floating-point number r between 0 and 1, that we call index indicator. These index indicators are assigned to node IDs, and encode which parallel edge to use when leaving the given node. Given two neighbouring nodes u, v and the number of parallel edges between them $l(u, v)$, the index of the chosen parallel edge can be calculated as $\lfloor r(u) * l(u, v) \rfloor + 1$. Any random value of index indicators can be decoded to an index of parallel edge that is available between two given nodes.

This way we do not need to access information about the structure of the multigraph when performing evolutionary operators, only when evaluating the solutions. We also apply the same method for direct representations in our experiments.

Direct Variable Length Representation for the Multigraph MSPP.
We adapt the representation found in [4] directly. The node IDs are integers and
thus a convenient way to store the parallel edge indicators that are floating-point
numbers below 1 is to add them to the node IDs. This way the chromosome is
the same length as in the simple graph problem. The integer parts indicate the
node sequence of the path and the fractional parts indicate which of the parallel
edges to use when leaving the nodes. The node sequence and sequence of parallel
edge indices can be decoded according to Algorithm 1.

The crossover, mutation and repair operators are the same as in [4], the only
modification is that crossovers can be conducted on any two parents that have
at least one node in common. In the simple graph case there is no point in
crossing two chromosomes that encode the same sequence of nodes, however for
the multigraph problem the overlap-based crossover can still be used to cross
the parallel edge indicators of the two solution paths. The mutation operator
generates new partial solutions by a random walk.

Algorithm 1. Decode direct variable length chromosome

Input: chromosome, G, v_O, v_D
Output: path: node sequence, indices: parallel edge indicator sequence
1 *path* ← Empty list+ v_O
2 *indices* ← Empty list
3 **for** i ← 2 **to** *length of chromosome* **do**
4 | *prevNode* ← Integer part of the $(i-1)$th gene of chromosome
5 | *nextNode* ← Integer part of the ith gene of chromosome
6 | *r(prevNode)* ← fractional part of the $(i-1)$th gene of chromosome
7 | *path* ← *path* + *nextNode*
8 | *indices* ← *indices* + ⌊ *r(prevNode)* * l(*prevNode,nextNode*) ⌋ + 1

9 **return** *path, indices*

An example of a chromosome from this representation that encodes the solution path in Fig. 1 is: [1.38, 6.82, 2.67, 4.51]. The value of the first gene encodes
that the first node in the solution path is 1 and the parallel edge to be used to
reach the second node, 6 is calculated as $\lfloor r(1) * l(1,6) \rfloor + 1 = \lfloor 0.38 * 3 \rfloor + 1 = 2$.

Direct Fixed Length Representation for the Multigraph MSPP. The
representation introduced in [16] is adapted. As before, we include parallel edge
indicators as fractional parts added to the genes that were previously integers.
The parallel edge indicator of node v is $r(v) = gene_v - \lfloor gene_v \rfloor$ where $gene_v$
is the vth gene of the chromosome. The node sequence and sequence of parallel
edge indices can be decoded according to Algorithm 2.

The uniform crossover is adapted without modification. There wasn't any
mutation specified, so we used the following mutation operator. When a chromosome is mutated each of its genes is reassigned randomly with probability 0.5.

This means that the integer part of the gene at locus i is changed to a random neighbour of the node i and the same fractional part is kept.

Algorithm 2. Decode direct fixed length chromosome

Input: chromosome, G, v_O, v_D

Output: path: node sequence, indices: parallel edge indicator sequence

1   $path \leftarrow$ Empty list $+ v_O$

2   $indices \leftarrow$ Empty list

3 **while** *Last node of path* $\neq v_D$ **do**

4      $prevNode \leftarrow$ Last node of *path*

5      $nextNode \leftarrow$ integer part of *prevNode*-th element of chromosome

6   **if** *nextNode is in path* **then**

7         $path \leftarrow path + v_D$ `// Path got stuck, will be penalised`

8   **else**

9         $r(prevNode) \leftarrow$ fractional part of the parallel edge indicator of node *prevNode*

10        $path \leftarrow path + nextNode$

11        $indices \leftarrow indices + \lfloor r(prevNode) * l(prevNode,nextNode) \rfloor + 1$

12 **return** *path, indices*

An example of a chromosome from this representation that encodes the solution path in Fig. 1 is: [6.38, 4.67, 2.24, 5.51, 4.09, 2.82].

Integer-Valued Priority Based Representation for the Multigraph MSPP. The representation used in [19] is extended. As before, we include parallel edge indicators as fractional parts added to the genes that were previously integers. The priority of node v can be found as $priority(v) = \lfloor gene_v \rfloor$, and the parallel edge indicator of node v is $r(v) = gene_v - \lfloor gene_v \rfloor$ where $gene_v$ is the vth gene of the chromosome. The node sequence and sequence of parallel edge indices can be decoded according to Algorithm 3. Insertion mutation and WMX is employed as in [19], without modifications.

An example of a chromosome from this representation that encodes the solution path in Fig. 1 is: [6.38, 3.67, 1.24, 2.51, 5.09, 4.82].

Random Key Based Representation for the Multigraph MSPP. The representation proposed in [14] is adapted. Here the chromosomes do not consist of integers for the simple graph MSPP. Thus here the parallel edge indicators are stored separately, making the genes two dimensional. The first value of the gene at locus i encodes the priority value of the node with ID i, while the second of the gene at locus i encodes the parallel edge indicator of the node with ID i. The sequence of nodes and parallel edge indices can be decoded from a chromosome according to Algorithm 3.

Algorithm 3. Decode priority based representations

Input: chromosome, G, v_O, v_D
Output: path: node sequence, indices: parallel edge indicator sequence
1   *path* ← Empty list + v_O
2   *indices* ← Empty list
3   **while** *Last element of path* $\neq v_D$ **do**
4   *neighbours* ← set of neighbours in G of last element in *path*
5   *allowed* ← set of elements in *neighbours* not in *path*
6   **if** *allowed is empty* **then**
7   *path* ← *path* + v_D // Path got stuck, will be penalised
8   **else**
         // Priority of node and parallel indicator of node is found
         differently for Integer priority and Random keys
9   *prevNode* ← Last node in path
10   *nextNode* ← node with maximum priority in *allowed*
11   *r(prevNode)* ← fractional part of the parallel edge indicator of node
         prevNode
12   *path* ← *path* + *nextNode*
13   *indices* ← *indices* + $\lfloor$ *r(prevNode)* * l(*prevNode,nextNode*) $\rfloor$ + 1

14 **return** *path, indices*

In [14] arithmetic crossover was used, but 1-point and 2-point crossovers are also mentioned as a possibility and we found in preliminary experiments that two-point crossover gives the best results. Thus we use two-point crossover and insertion mutation. Both the mutation and crossover mechanism is independent of the values of the genes and thus is straightforward to apply on two dimensional genes.

An example of a chromosome from this representation that encodes the solution path in Fig. 1 is: [(0.93,0.38), (0.36,0.67), (0.12,0.24), (0.25,0.51), (0.51,0.09), (0.45,0.82)].

Additional Mutation Operator. We define an additional mutation operator used with all four representations to change some of the parallel edge indicators. This mutation cannot make a solution path infeasible, because that only depends on the encoded node sequence. For this reason the additional mutation operator is used with probability 1 on all candidates, however it might not always introduce any changes. For a given chromosome each of its parallel edge indicators are reassigned randomly with a small probability, which in the numerical experiments is set to 0.05.

4.2 Heuristic Initialisation

Inspired by previous works in the literature that aim to avoid detours, we propose a novel heuristic initialisation technique that can also be used with priority based genetic representations. Compared to the method proposed in [23], our approach similarly aims to reduce the chance of backwards moves, but not to eliminate them. We incorporate the roles of IDs and priorities into a single priority value assigned to each node in a quasi-random way.

Solutions encoded by random key representation can be transformed to any of the other three representations, while this is not true the other way around. We introduce the initialisation technique for random key representation, and later describe how to use it for the other ones.

The idea is to give higher priorities to nodes closer to the destination node more often than to the ones far from it, thereby discouraging detours. The method incorporates knowledge about the network structure and randomisation to provide a diverse initial population of high quality. The priority p is assigned to node v according to Eq. (2). The hopcount (distance without any weights) of node v from the destination node v_D in the network G is denoted $h(v, v_D)$, and τ_{max} denotes a randomisation coefficient.

$$p(v, v_D) = -h(v, v_D, G) + \tau, \quad \tau \in (0, \tau_{max}) \tag{2}$$

The likelihood of detours appearing in the decoded paths can be controlled by the parameter τ_{max}. The higher τ_{max} is, the more random the priorities are, and the less prominent is the effect of the heuristic initialisation compared to a purely random one. If $\tau_{max} < 1$, all initial solutions will be paths without detours, with minimal hop-counts, which is undesirable when the costs are highly inhomogeneous and higher hopcount routes might still be efficient. If $1 < \tau_{max} < 2$ small detours are possible, an encoded path might move from a node to another one with the same hopcount, as depicted in Fig. 2. If $2 < \tau_{max}$ moving to a node that is at higher hopcount from the destination is possible and becomes more probable with the increase of τ_{max}.

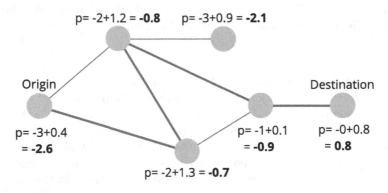

Fig. 2. Illustration of the role of τ_{max} in heuristic initialisation. Here $\tau_{max} = 1.5$, and thus detours are possible as demonstrated.

The resulting values need to be transformed according to the representations before they are fed to the algorithms as initial populations. For random key encoding, they need to be normalised to fit the appropriate intervals. For integer valued priority encoding the nodes are converted to integers by sorting them into increasing order according to their priorities and assign to each node its rank. For direct encodings the priorities are converted to node-based representations by decoding them as usual, each node is followed by its neighbour with the highest priority. This way the method can be used with all four representations. The parallel edge indicators are assigned purely randomly.

A crucial point is that the heuristic initialisation method should be easily computable compared to the original problem being solved. The proposed method makes use of the hopcount of each node in the graph from the destination node, which can be computed in $O(V + E)$ time. This is significantly lower than solving the multigraph MSPP, which is in general NP-hard.

5   Results

The proposed algorithms are tested empirically and their performances are evaluated compared to the true Pareto front using a set of performance indicators. The true Pareto front was found by a state-of-the-art exact algorithm NAMOA* [22], that was adapted to the multigraph problem.

5.1   Implementation Details

We use the NSGA-II [8] algorithm to compare the representation and initialisation techniques. The selection and elitism mechanisms are defined by NSGA-II. In all experiments the population size was set to n, the number of nodes, as it is often done in the literature. The maximum number of generations was set to 400. In all reported results the crossover probability is 1 and the mutation probability is 0.15. All numerical tests are performed on Queen Mary's Apocrita HPC facility [2]. The methods are implemented in Python, for the NSGA-II implementation the inspyred package [1] was used.

The fitness of a valid path is calculated according to Eq. (1). In all cases it might happen that some candidates do not correspond to a feasible path, because the decoded path does not reach the destination node. In these cases a penalty function is used, that assigns a large cost to such candidates. The penalty is larger the further away the path ends from the destination node, measured by hopcount. The fitness of an infeasible path P' that does not reach the destination node is calculated according to Eq. (3), where $\overline{cost_{max}}$ is the k-dimensional vector where each component equals the maximum value of any cost component in the given instance.

$$C(P') = \sum_{e \in P'} \overline{cost(e)} + \overline{cost_{max}} * h(P', v_D) \tag{3}$$

5.2   Test Instances

The algorithms are evaluated using 8 test instances for the 2 objective problem and 6 test instances for the 3 objective problem. The test instances are described in Table 1.

Table 1. The multigraph MSPP instances used for evaluation. Each instance has two variants one for 2 objectives and one for 3 objectives with the same graph structure.

Instance ID	Number of nodes	Network type	l_{max}	Correlation between objectives
I_1	49	Waxman	5	Negative
I_2	49	Waxman	10	Negative
I_3	100	Waxman	5	Negative
I_4	100	Waxman	10	Negative
I_5	49	Grid	5	Negative
I_6	49	Grid	10	Negative
I_7	100	Grid	5	Negative
I_8	100	Grid	10	Negative

We use Waxman networks [26] on 49 and 100 nodes and square grid networks with the same number of nodes (7 by 7 and 10 by 10 nodes). Each edge in these simple graphs is converted to a multi-edge by assigning a cost matrix to it. The number of parallel edges (number of rows of the cost matrix) is randomly chosen between 1 and l_{max}, where l_{max} the maximum allowed number of parallel edges, 5 or 10 in this case. All parallel edges between the same two nodes have non-dominated cost-vectors. The cost assignment method described in [21] was used to generate costs with negative correlation, because this is the case where a multi-objective approach is essential for real world applications and also these kind of instances are the most challenging for the exact algorithms. The cost components are from the interval $(1, 1000)$. The origin and destination nodes are specified as two endpoints of a diameter of the network, to ensure that they are not too close that could result in a trivial problem.

For 3 objectives I_8 and I_7 the exact algorithm did not terminate in five days, we failed to find the true Pareto front and thus those instances are left out of the investigation.

5.3   Comparing Representations with Purely Random Initialisation

The solution quality of approximate algorithms can be evaluated in different ways. We compare them to the true Pareto front by dominance-compliant quality indicators recommended in [10], the hypervolume indicator, the epsilon indicator and the R_3 indicator. For the calculation of the R_3 and hypervolume metric, the fitnesses of the solutions are normalised, such that the components of the fitnesses of the real Pareto fronts are between 0 and 1. This means that the fitness

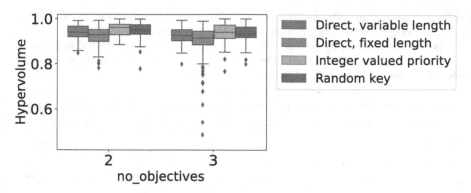

Fig. 3. The hypervolume indicator for 50 runs of the algorithms with purely random initialisation on 8 test instances for the two objective multigraph MSPP, and 6 test instances for the three objective multigraph MSPP.

components of the approximate solutions might be above 1. The reference point for the hypervolume calculation is set to 1.5 for all objectives and approximate solutions that do not dominate this reference point are ignored. The hypervolume defined by the approximate solutions is divided by the hypervolume defined by the true Pareto front to get a value between 0 and 1.

The performance achieved by the different representations as measured by the hypervolume indicator is presented in Fig. 3. Both in the two and three objective cases the two priority-based representations perform the best, with the integer valued priority showing a slightly higher average. The worst performing representation is the direct fixed-length one, which agrees with our expectations based on the literature. The results are similar for the other performance indicators. The Wilcoxon signed rank test is performed for all pairs of the representations for both the bi-objective and three objective problems. The only case where the difference was non-significant (with p-value $= 0.05$) is the two priority-based representations in the three-objective case for all three performance indicators (epsilon, R3, hypervolume). In all other cases, there are significant differences between the representations.

5.4 Comparing Initialisation Techniques

The same representations were also tested using the proposed heuristic initialisation method. Only the initial population is changed, the genetic operators and parameters are the same. The resulting solution qualities after $401 * n$ function evaluations are described in Table 2. All representations perform better according to all performance metrics when heuristic initialisation is used compared to the purely random initialisation. This difference was found significant in case of all representations by the Wilcoxon signed rank test. The best performing algorithm is the random-keys based one with heuristic initialisation, according to Table 2, however the difference between the performances of the representations is less prominent when the heuristic initialisation is used.

Table 2. The average values of the three performance indicators for 50 runs of the algorithms on each of the 8 test instances with the 4 main representations with the two initialisation methods for the two and three objective multigraph MSPP. For heuristic initialisation $\tau_{max} = 1.5$ was used. The best values for the two and three objective cases separately are indicated in bold for each instance.

Representation	Direct, variable l.		Direct, fixed l.		Integer priority		Random key	
Initialisation	Rand.	Heur.	Rand.	Heur.	Rand.	Heur.	Rand.	Heur.
No objectives								
Epsilon indicator								
2	1.779	1.574	1.853	1.453	1.574	1.455	1.641	**1.427**
3	1.453	1.337	1.577	1.237	1.376	1.220	1.377	**1.188**
R3 indicator								
2	0.017	0.010	0.022	0.007	0.011	0.007	0.013	**0.006**
3	0.013	0.007	0.021	0.004	0.010	0.004	0.010	**0.003**
Hypervolume indicator								
2	0.941	0.959	0.926	0.970	0.955	0.969	0.952	**0.972**
3	0.927	0.948	0.907	0.970	0.942	0.967	0.938	**0.974**

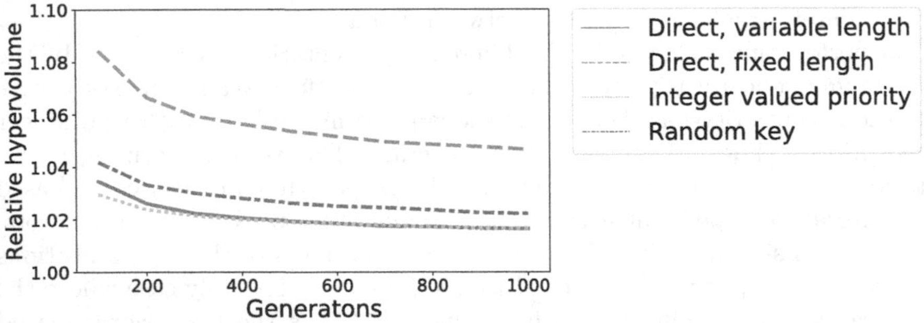

Fig. 4. The ratio of the average hypervolume achieved by the heuristic and random initialisations as a function of the elapsed generations. A value above 1 means that the heuristic initialisation ($\tau_{max} = 1.5$) achieved a higher hypervolume.

A heuristic initialisation method should allow the evolutionary process to start with a higher quality population and also should not lead to premature convergence. To illustrate these properties we visualise the ratio of the hypervolume achieved with the heuristic and purely random initialisations throughout 1000 generations in Fig. 4. The ratio stays above 1 through the evolutionary process, which means that heuristic initialisation leads to better solution quality at all stages for all the representations. The improvement is the most prominent in the earlier generations. Thus we can conclude that speed-up is achieved and the premature convergence is avoided.

Note that the direct fixed length representation profits the most from the heuristic initialisation technique. In fact it is competitive with the priority based representations when used with the heuristic initialisation (see Table 2), while it is significantly worse than others when used with the purely random initialisation.

6 Conclusion and Future Work

Four previously proposed representations for the MSPP were extended to the multigraph case and the performances of the resulting algorithms are compared according to three performance indicators. The extension is done in a way that it can be used for multigraph networks with inhomogeneous numbers of parallel edges between pairs of nodes, even with priority-based representations. We observed that the best performing representations are the priority-based representations in almost all cases.

A heuristic initialisation technique was proposed that can be used with all four representations. It was found that the heuristic initialisation improves the solution quality significantly for three of the four representations, the direct fixed length representation, the integer-valued priority representation and the random key-based representation.

Future work includes refining the representation of parallel edges and decreasing the number of possible chromosomes that encode the same solution path, which could potentially make the algorithms more efficient. Another future direction is the extension to constrained problems particularly the investigation and design of bespoke crossover and mutation operators. The extension will allow the proposed approach to be tested on real-world problems and to be compared with other exact and approximate algorithms.

References

1. Inspyred: Bio-inspired algorithms in Python. https://pythonhosted.org/inspyred/. Accessed 30 Oct 2019
2. This research utilised Queen Mary's apocrita HPC facility, supported by QMUL research-it. http://doi.org/10.5281/zenodo.438045. Accessed 12 Nov 2019
3. Abbaspour, R.A., Samadzadegan, F.: An evolutionary solution for multimodal shortest path problem in metropolises. Comput. Sci. Inf. Syst. 7(4), 789–811 (2010)
4. Ahn, C.W., Ramakrishna, R.S.: A genetic algorithm for shortest path routing problem and the sizing of populations. IEEE Trans. Evol. Comput. 6(6), 566–579 (2002)
5. Bean, J.C.: Genetic algorithms and random keys for sequencing and optimization. ORSA J. Comput. 6(2), 154–160 (1994)
6. Chen, J., et al.: Toward a more realistic, cost-effective, and greener ground movement through active routing: a multiobjective shortest path approach. IEEE Trans. Intell. Transp. Syst. 17(12), 3524–3540 (2016)
7. Chitra, C., Subbaraj, P.: A nondominated sorting genetic algorithm solution for shortest path routing problem in computer networks. Expert Syst. Appl. 39(1), 1518–1525 (2012)
8. Deb, K., Pratap, A., Agarwal, S., Meyarivan, T.: A fast and elitist multiobjective genetic algorithm: NSGA-II. IEEE Trans. Evol. Comput. 6(2), 182–197 (2002)

9. Ehmke, J.F., Campbell, A.M., Thomas, B.W.: Vehicle routing to minimize time-dependent emissions in urban areas. Eur. J. Oper. Res. **251**(2), 478–494 (2016)
10. Fonseca, C.M., Knowles, J.D., Thiele, L., Zitzler, E.: A tutorial on the performance assessment of stochastic multiobjective optimizers. In: Third International Conference on Evolutionary Multi-Criterion Optimization (EMO 2005), vol. 216, p. 240 (2005)
11. Garaix, T., Artigues, C., Feillet, D., Josselin, D.: Vehicle routing problems with alternative paths: an application to on-demand transportation. Eur. J. Oper. Res. **204**(1), 62–75 (2010)
12. Gen, M., Altiparmak, F., Lin, L.: A genetic algorithm for two-stage transportation problem using priority-based encoding. OR Spectr. **28**(3), 337–354 (2006)
13. Gen, M., Cheng, R., Wang, D.: Genetic algorithms for solving shortest path problems. In: Proceedings of 1997 IEEE International Conference on Evolutionary Computation (ICEC 1997), pp. 401–406. IEEE (1997)
14. Gen, M., Lin, L.: A new approach for shortest path routing problem by random key-based GA. In: Proceedings of the 8th Annual Conference on Genetic and Evolutionary Computation, pp. 1411–1412. ACM (2006)
15. Hrnčíř, J., Rovatsos, M., Jakob, M.: Ridesharing on timetabled transport services: a multiagent planning approach. J. Intell. Transp. Syst. **19**(1), 89–105 (2015)
16. Inagaki, J., Haseyama, M., Kitajima, H.: A genetic algorithm for determining multiple routes and its applications. In: ISCAS 1999, Proceedings of the 1999 IEEE International Symposium on Circuits and Systems VLSI (Cat. No. 99CH36349), vol. 6, pp. 137–140. IEEE (1999)
17. Ji, Z., Kim, Y.S., Chen, A.: Multi-objective α-reliable path finding in stochastic networks with correlated link costs: a simulation-based multi-objective genetic algorithm approach (SMOGA). Expert Syst. Appl. **38**(3), 1515–1528 (2011)
18. Li, R., Leung, Y., Huang, B., Lin, H.: A genetic algorithm for multiobjective dangerous goods route planning. Int. J. Geogr. Inf. Sci. **27**(6), 1073–1089 (2013)
19. Lin, L., Gen, M.: An effective evolutionary approach for bicriteria shortest path routing problems. IEEJ Trans. Electron. Inf. Syst. **128**(3), 416–423 (2008)
20. Lin, L., Gen, M.: Priority-based genetic algorithm for shortest path routing problem in OSPF. In: Gen, M., et al. (eds.) Intelligent and Evolutionary Systems, pp. 91–103. Springer, Heidelberg (2009). https://doi.org/10.1007/978-3-540-95978-6_7
21. Machuca, E., Mandow, L., de la Cruz, J.L.P., Ruiz-Sepulveda, A.: An empirical comparison of some multiobjective graph search algorithms. In: Dillmann, R., Beyerer, J., Hanebeck, U.D., Schultz, T. (eds.) KI 2010. LNCS (LNAI), vol. 6359, pp. 238–245. Springer, Heidelberg (2010). https://doi.org/10.1007/978-3-642-16111-7_27
22. Mandow, L., De la Cruz, J.P., et al.: A new approach to multiobjective A* search. In: IJCAI, vol. 8. Citeseer (2005)
23. Mohemmed, A.W., Sahoo, N.C., Geok, T.K.: Solving shortest path problem using particle swarm optimization. Appl. Soft Comput. **8**(4), 1643–1653 (2008)
24. Munetomo, M., Takai, Y., Sato, Y.: An adaptive network routing algorithm employing path genetic operators. In: ICGA (1997)
25. Qian, J., Eglese, R.: Fuel emissions optimization in vehicle routing problems with time-varying speeds. Eur. J. Oper. Res. **248**(3), 840–848 (2016)
26. Waxman, B.M.: Routing of multipoint connections. IEEE J. Sel. Areas Commun. **6**(9), 1617–1622 (1988)
27. Yu, H., Lu, F.: A multi-modal route planning approach with an improved genetic algorithm. Adv. Geo-Spat. Inf. Sci. **193**, 343–348 (2012)

The Univariate Marginal Distribution Algorithm Copes Well with Deception and Epistasis

Benjamin Doerr[1] and Martin S. Krejca[2(✉)]

[1] Laboratoire d'Informatique (LIX), CNRS, École Polytechnique,
Institute Polytechnique de Paris, Palaiseau, France
`doerr@lix.polytechnique.fr`
[2] Hasso Plattner Institute, University of Potsdam, Potsdam, Germany
`martin.krejca@hpi.de`

Abstract. In their recent work, Lehre and Nguyen (FOGA 2019) show that the univariate marginal distribution algorithm (UMDA) needs time exponential in the parent populations size to optimize the DECEIVING-LEADINGBLOCKS (DLB) problem. They conclude from this result that univariate EDAs have difficulties with deception and epistasis.

In this work, we show that this negative finding is caused by an unfortunate choice of the parameters of the UMDA. When the population sizes are chosen large enough to prevent genetic drift, then the UMDA optimizes the DLB problem with high probability with at most $\lambda(\frac{n}{2} + 2e \ln n)$ fitness evaluations. Since an offspring population size λ of order $n \log n$ can prevent genetic drift, the UMDA can solve the DLB problem with $O(n^2 \log n)$ fitness evaluations. In contrast, for classic evolutionary algorithms no better run time guarantee than $O(n^3)$ is known, so our result rather suggests that the UMDA can cope well with deception and epistatis.

Together with the result of Lehre and Nguyen, our result for the first time rigorously proves that running EDAs in the regime with genetic drift can lead to drastic performance losses.

Keywords: Estimation-of-distribution algorithm · Univariate marginal distribution algorithm · Run time analysis · Epistasis · Theory

1 Introduction

Estimation-of-distribution algorithms (EDAs) are randomized search heuristics that evolve a probabilistic model of the search space in an iterative manner. Starting with the uniform distribution, an EDA takes samples from its current model and then adjusts it such that better solutions have a higher probability of being generated in the next iteration. This method of refinement leads to gradually better solutions and performs well on many practical problems, often outperforming competing approaches [21].

© Springer Nature Switzerland AG 2020
L. Paquete and C. Zarges (Eds.): EvoCOP 2020, LNCS 12102, pp. 51–66, 2020.
https://doi.org/10.1007/978-3-030-43680-3_4

Theoretical analyses of EDAs also often suggest an advantage of EDAs when compared to evolutionary algorithms (EAs); for an in-depth survey of run time results for EDAs, please refer to the article by Krejca and Witt [16]. With respect to simple unimodal functions, EDAs seem to be comparable to EAs. For example, Sudholt and Witt [22] proved that the two EDAs cGA and 2-MMAS$_{ib}$ have an expected run time of $\Theta(n \log n)$ on the standard theory benchmark function ONEMAX (assuming optimal parameter settings; n being the problem size), which is a run time that many EAs share. The same is true for the EDA UMDA, as shown by the results of Krejca and Witt [15], Lehre and Nguyen [17], and Witt [24]. For the benchmark function LEADINGONES, Dang and Lehre [4] proved an expected run time of $O(n^2)$ for the UMDA when setting the parameters right, which is, again, a common run time bound for EAs on this function. One result suggesting that EDAs can outperform EAs on unimodal function was given by Doerr and Krejca [8]. They proposed an EDA called sig-cGA, which has an expected run time of $O(n \log n)$ on both ONEMAX and LEADINGONES – a performance not known for any classic EA or EDA.

For the class of all linear functions, the EDAs perform slightly worse than EAs. The classical $(1 + 1)$ evolutionary algorithm optimizes all linear functions in time $O(n \log n)$ in expectation [12]. In contrast, the conjecture of Droste [11] that the cGA does not perform equally well on all linear functions was recently proven by Witt [23], who showed that the cGA has an $\Omega(n^2)$ expected run time on the binary value function. We note that the binary value function was found harder also for classical EAs. While the $(1 + \lambda)$ evolutionary algorithm optimizes ONEMAX with $\Theta(n\lambda \log \log \lambda / \log \lambda)$ fitness evaluations, it takes $\Theta(n\lambda)$ fitness evaluations for the binary value functions [9].

For the bimodal JUMP$_k$ benchmark function, which has a local optimum with a Hamming distance of k away from the global optimum, EDAs seem to drastically outperform EAs. Hasenhörl and Sutton [13] recently proved that the cGA only has a run time of $\exp\big(O(k + \log n)\big)$. In contrast, common EAs have a run time of $\Theta(n^k)$ [12] (mutation-only) or $\Theta(n^{k-1})$ [3] (mutation and crossover), and only get to $O(n \log n + kn + 4^k)$ by using crossover in combination with diversity mechanisms like Island models [2]. Doerr [6] proved that the cGA even has an expected run time of $O(n \log n)$ on JUMP$_k$ if $k < \frac{1}{20} \ln n$, meaning that the cGA is unfazed by the gap of k separating the local from the global optimum.

Another result in favor of EDAs was given by Chen et al. [1], who introduced the SUBSTRING function and proved that the UMDA optimizes it in polynomial time, whereas the $(1 + 1)$ evolutionary algorithm has an exponential run time, both with high probability. In the SUBSTRING function, only substrings of length αn, for $\alpha \in (0, 1)$, of the global optimum are relevant to the fitness of a solution, and these substrings provide a gradient to the optimum. In the process, the $(1+1)$ evolutionary algorithm loses bits that are not relevant anymore for following the gradient (but relevant for the optimum). The UMDA fixes its model for correct positions while it is following the gradient and thus does not lose these bits.

The first, and so far only, result to suggest that EDAs can be drastically worse than EAs was recently stated by Lehre and Nguyen [18] via the DECEIVINGLEA-

DINGBLOCKS function (DLB for short), which they introduce and which consists of blocks of size 2, each with a deceiving function value, that need to be solved sequentially. The authors prove that many common EAs optimize DLB within $O(n^3)$ fitness evaluations in expectation, whereas the UMDA has a run time of $e^{\Omega(\mu)}$ (where μ is an algorithm-specific parameter that often is chosen as a small power of n) for a large regime of parameters. Only for extreme parameter values, the authors prove an expected run time of $O(n^3)$ also for the UMDA.

In this paper, we prove that the UMDA is, in fact, able to optimize DLB in time $O(n^2 \log n)$ with high probability if its parameters are chosen more carefully (Theorem 5). Note that our result is better than any of the run times proven in the paper by Lehre and Nguyen [18]. We achieve this run time by choosing the parameters of the UMDA such that its model is unlikely to degenerate during the run time (Lemma 6). Here by *degenerate* we mean that the sampling frequencies approach the boundary values 0 and 1 without that this is justified by the objective function. This leads to a probabilistic model that is strongly concentrated around a single search point. This effect is often called *genetic drift* [22]. While it appears natural to choose the parameters of an EDA as to prevent genetic drift, it also has been proven that genetic drift can lead to a complicated run time landscape and inferior performance (see Lengler et al. for the cGA [19]).

In contrast to our setting, for their exponential lower bound, Lehre and Nguyen [18] use parameters that lead to genetic drift. Once the probabilistic model it is sufficiently degenerated, the progress of the UMDA is so slow that even to leave the local optima of DLB (that have a better search point in Hamming distance two only), the EDA takes time exponential in μ.

Since the UMDA shows a good performance in the (more natural) regime without genetic drift and was shown inferior only in the regime with genetic drift, we disagree with the statement of Lehre and Nguyen [18] that there are "inherent limitations of univariate EDAs against deception and epistasis".

In addition to the improved run time, we derive our result using only tools commonly used in the analysis of EDAs and EAs, whereas the proof of the polynomial run time of $O(n^3)$ for the UMDA with uncommon parameter settings [18] uses the level-based population method, which is an advanced tool that can be hard to use. We are thus optimistic that our analysis method can be useful also in other run time analyses of EDAs.

Last, we complement our theoretical result with an empirical comparison of the UMDA to two other evolutionary algorithms. The outcome of these experiments suggests that the UMDA outperforms the competing approaches while also having a smaller variance.

The remainder of this paper is structured as follows: in Sect. 2, we introduce our notation, formally define DLB and the UMDA, and we state the tools we use in our analysis. Section 3 contains our main result (Theorem 5) and discusses its proof informally before stating the different lemmas used to prove it. In Sect. 4, we discuss our empirical results. Last, we conclude this paper in Sect. 5.

2 Preliminaries

We are concerned with the run time analysis of algorithms optimizing pseudo-Boolean functions, that is, functions $f\colon \{0,1\}^n \to \mathbb{R}$, where $n \in \mathbb{N}$ denotes the dimension of the problem. Given a pseudo-Boolean function f and a bit string x, we refer to f as a *fitness function*, to x as an *individual*, and to $f(x)$ as the *fitness of x*.

For $n_1, n_2 \in \mathbb{N} := \{0,1,2,\ldots\}$, we define $[n_1..n_2] = [n_1, n_2] \cap \mathbb{N}$, and for an $n \in \mathbb{N}$, we define $[n] = [1..n]$. From now on, if not stated otherwise, the variable n always denotes the problem size. For a vector x of length n, we denote its component at index $i \in [n]$ by x_i and, for and index set $I \subseteq [n]$, we denote the subvector of length $|I|$ consisting only of the components at indices in I by x_I. Further, let $|x|_1$ denote the number of 1s of x and $|x|_0$ its number of 0s.

DeceivingLeadingBlocks. The pseudo-Boolean function DECEIVINGLEA-DINGBLOCKS (abbreviated as DLB) was introduced by Lehre and Nguyen [18] as a deceptive version of the well known benchmark function LEADINGONES. In DLB, an individual x of length n is divided into blocks of equal size 2. Each block consists of a trap, where the fitness of each block is determined by the number of 0s (minus 1), except that a block of all 1s has the best fitness of 2. The overall fitness of x is then determined by the longest prefix of blocks with fitness 2 plus the fitness of the following block. Note that in order for the chunking of DLB to make sense, it needs to hold that 2 divides n. In the following, we always assume this implicitly.

We now provide a formal definition of DLB. To this end, we first introduce the function DECEIVINGBLOCK $\colon \{0,1\}^2 \to [0..2]$ (abbreviated as DB), which determines the fitness of a block (of size 2). For all $x \in \{0,1\}^2$, we have

$$\mathrm{DB}(x) = \begin{cases} 2 & \text{if } |x|_1 = 2, \\ |x|_0 - 1 & \text{else.} \end{cases}$$

Further, we define the function PREFIX $\colon \{0,1\}^n \to [0..n]$, which determines the longest prefix of x with blocks of fitness 2. For a logic formula P, let $[P]$ be 1 if P is true and 0 otherwise. We define, for all $x \in \{0,1\}^n$,

$$\mathrm{PREFIX}(x) = \sum_{i=1}^{n/2} \big[\forall j \le i \colon \mathrm{DB}(x_{\{2i-1,2i\}}) = 2 \big].$$

DLB is now defined as follows for all $x \in \{0,1\}^n$:

$$\mathrm{DLB}(x) = \begin{cases} n & \text{if } \mathrm{PREFIX}(x) = n, \\ \sum_{i=1}^{\mathrm{PREFIX}(x)+1} \mathrm{DB}(x_{\{2i-1,2i\}}) & \text{else.} \end{cases}$$

The Univariate Marginal Distribution Algorithm. Our algorithm of interest is the UMDA ([20]; Algorithm 1) with parameters $\mu, \lambda \in \mathbb{N}^+$, $\mu \leq \lambda$. It maintains a vector p *(frequency vector)* of probabilities *(frequencies)* of length n as its probabilistic model. This vector is used to sample an individual $x \in \{0,1\}^n$, which we denote as $x \sim \text{sample}(p)$, such that, for all $y \in \{0,1\}^n$,

$$\Pr[x = y] = \prod_{\substack{i=1; \\ y_i=1}}^{n} p_i \prod_{\substack{i=1; \\ y_i=0}}^{n} (1 - p_i).$$

The UMDA updates this vector iteratively in the following way: first, λ individuals are sampled. Then, among these λ individuals, a subset of μ with the highest fitness is chosen (breaking ties uniformly at random), and, for each index $i \in [n]$, the frequency p_i is set to the relative number of 1s at position i among the μ best individuals. Last, if a frequency p_i is below $\frac{1}{n}$, it is increased to $\frac{1}{n}$, and, analogously, frequencies above $1 - \frac{1}{n}$ are set to $1 - \frac{1}{n}$. Capping into the interval $[\frac{1}{n}, 1 - \frac{1}{n}]$ circumvents frequencies from being stuck at the extremal values 0 or 1. Last, we denote the frequency vector of iteration $t \in \mathbb{N}$ with $p^{(t)}$.

Algorithm 1. The UMDA [20] with parameters μ and λ, $\mu \leq \lambda$, maximizing a fitness function $f \colon \{0,1\}^n \to \mathbb{R}$ with $n \geq 2$

1 $t \leftarrow 0$;
2 $p^{(t)} \leftarrow (\frac{1}{2})_{i \in [n]}$;
3 **repeat** ▷ *iteration t*
4   **for** $i \in [\lambda]$ **do** $x^{(i)} \sim \text{sample}(p^{(t)})$;
5   let $y^{(1)}, \ldots, y^{(\mu)}$ denote the μ best individuals out of $x^{(1)}, \ldots, x^{(\lambda)}$ (breaking ties uniformly at random);
6   **for** $i \in [n]$ **do** $p_i^{(t+1)} \leftarrow \frac{1}{\mu} \sum_{j=1}^{\mu} y_i^{(j)}$;
7   restrict $p^{(t+1)}$ to the interval $[\frac{1}{n}, 1 - \frac{1}{n}]$;
8      $t \leftarrow t + 1$;
9 **until** termination criterion met;

Run Time Analysis. When analyzing the run time of the UMDA optimizing a fitness function f, we are interested in the number T of fitness function evaluations until an optimum of f is sampled for the first time. Since the UMDA is a randomized algorithm, this run time T is a random variable. Note that the run time of the UMDA is at most λ times the number of iterations until an optimum is sampled for the first time, and it is at least $(\lambda - 1)$ times this number.

In the area of run time analysis of randomized search heuristics, it is common to give bounds for the expected value of the run time of the algorithm under investigation. This is uncritical when the run time is concentrated around its expectation, as often observed for classical evolutionary algorithms. For EDAs, it has been argued, among others in [6], that it is preferable to give bounds that hold with high probability. This is what we shall aim at in this work as well. Of course, it would be even better to give estimates in a distributional sense, as argued for in [5], but this appears to be difficult for EDAs, among others, because of the very different behavior in the regimes with and without strong genetic drift.

Probabilistic Tools. We use the following results in our analysis.
In order to prove statements on random variables that hold with high probability, we use the following commonly known Chernoff bound.

Theorem 1 (Chernoff bound [7, Theorem 10.5], [14]**).** *Let $k \in \mathbb{N}$, $\delta \in [0,1]$, and let X be the sum of k independent random variables, each taking values in $[0,1]$. Then*

$$\Pr\big[X \leq (1-\delta)\mathrm{E}[X]\big] \leq \exp\left(-\frac{\delta^2 \mathrm{E}[X]}{2}\right).$$

The next lemma tells us that, for a random X following a binomial law, the probability of exceeding $\mathrm{E}[X]$ is bounded from above by roughly the term with the highest probability.

Lemma 2 ([7, Eq. (10.62)]**).** *Let $k \in \mathbb{N}$, $p \in [0,1]$, $X \sim \mathrm{Bin}(k,p)$, and let $m \in \big[\mathrm{E}[X]+1..k\big]$. Then*

$$\Pr[X \geq m] \leq \frac{m(1-p)}{m - \mathrm{E}[X]} \cdot \Pr[X = m].$$

We use Lemma 2 for the converse case, that is, in order to bound the probability that a binomially distributed random variable is smaller than its expected value.

Corollary 3. *Let $k \in \mathbb{N}$, $p \in [0,1]$, $X \sim \mathrm{Bin}(k,p)$, and let $m \in \big[0..\mathrm{E}[X]-1\big]$. Then*

$$\Pr[X \leq m] \leq \frac{(k-m)p}{\mathrm{E}[X] - m} \cdot \Pr[X = m].$$

Proof. Let $\overline{X} := k - X$, and let $\overline{m} := k - m$. Note that $\overline{X} \sim \mathrm{Bin}(k, 1-p)$ with $\mathrm{E}[\overline{X}] = k - \mathrm{E}[X]$ and that $\overline{m} \in \big[\mathrm{E}[\overline{X}]+1..k\big]$. With Lemma 2, we compute

$$\Pr[X \leq m] = \Pr[\overline{X} \geq \overline{m}] \leq \frac{\overline{m}p}{\overline{m} - \mathrm{E}[\overline{X}]} \cdot \Pr[\overline{X} = \overline{m}] = \frac{(k-m)p}{\mathrm{E}[X] - m} \cdot \Pr[X = m],$$

which proves the claim. □

Last, the following theorem deals with a *neutral* bit in a fitness function f, that is, a position $i \in [n]$ such that bit values at i do not contribute to the fitness value at all. The theorem (from [10], extending a similar result [25, Theorem 4.5]) states that if the UMDA optimizes such an f, then the frequency at position i stays close to its initial value $\frac{1}{2}$ for $\Omega(\mu)$ iterations. We go more into detail about how this relates to DLB at the beginning of Sect. 3.

Theorem 4 ([10, Corollary 2]). *Consider the UMDA optimizing a fitness function f with a neutral bit $i \in [n]$. Then, for all $d > 0$ and all $t \in \mathbb{N}$, we have*

$$\Pr\left[\forall t' \in [0..t]\colon |p_i^{(t')} - \tfrac{1}{2}| < d\right] \geq 1 - 2\exp\left(-\frac{d^2\mu}{2t}\right).$$

3   Run Time Analysis

In the following, we prove that the UMDA optimizes DECEIVINGLEADING-BLOCKS efficiently, which is the following theorem.

Theorem 5. *Let $c_\mu, c_\lambda \in (0,1)$ be constants chosen sufficiently large or small, respectively. Consider the UMDA optimizing DECEIVINGLEADINGBLOCKS with $\mu \geq c_\mu n \ln n$ and $\mu/\lambda \leq c_\lambda$. Then the UMDA samples the optimum after $\lambda(\frac{n}{2} + 2e \ln n)$ fitness function evaluations with a probability of at least $1 - 9n^{-1}$.*

Before we present the proof, we sketch its main ideas and introduce important notation. We show that the frequencies of the UMDA are set to $1 - \frac{1}{n}$ block-wise from left to right with high probability. We formalize this concept by defining that a block $i \in [\frac{n}{2}]$ is *critical* (in iteration t) if and only if $p_{2i-1}^{(t)} + p_{2i}^{(t)} < 2 - \frac{2}{n}$ and, for each index $j \in [2i - 2]$, the frequency $p_j^{(t+1)}$ is at $1 - \frac{1}{n}$. Intuitively, a critical block is the first block whose frequencies are not at their maximum value. We prove that a critical block is optimized within a single iteration with high probability if we assume that its frequencies are not below $(1 - \varepsilon)/2$, for $\varepsilon \in (0,1)$ being a constant.

In order to assure that the frequencies of each block are at least $(1 - \varepsilon)/2$ until it becomes critical, we show that most of the frequencies right of the critical block are not impacted by the fitness function. We call such frequencies *neutral*. More formally, a frequency p_i is neutral in iteration t if and only if the probability to have a 1 at position i in each of the μ *selected* individuals equals $p_i^{(t)}$. Note that since we assume that $\mu = \Omega(n \log n)$, the impact of the genetic drift on neutral frequencies is low with high probability (Theorem 4).

We know which frequencies are neutral and which are not by the following key observation: consider a population of λ individuals of the UMDA during iteration t; only the first (leftmost) block that has strictly fewer than μ 11s is relevant for selection, since the fitness of individuals that do not have a 11 in this block cannot be changed by bits to the right anymore. We call this block *selection-relevant*. Note that this is a notion that depends on the random offspring population in iteration t, whereas the notion *critical* depends only on $p^{(t)}$.

The consequences of a selection-relevant block are as follows: if block $i \in [\frac{n}{2}]$ is selection-relevant, then all frequencies in blocks left of i are set to $1 - \frac{1}{n}$, since there are at least μ individuals with 11s. All blocks right of i have *no* impact on the selection process: if an individual has no 11 in block i, its fitness is already fully determined by all of its bits up to block i by the definition of DLB. If an individual has a 11 in block i, it is definitely chosen during selection, since there are fewer than μ such individuals and since its fitness is better than that of all of the other individuals that do not have a 11 in block i. Thus, its bits at positions in blocks right of i are irrelevant for selection. Overall, since the bits in blocks right of i do not matter, the frequencies right of block i get no signal from the fitness function and are thus neutral (Lemma 6).

Regarding block i itself, all of the individuals with 11s are chosen, since they have the best fitness. Nonetheless, individuals with a 00, 01, or 10 can also be chosen, where an individual with a 00 in block i is preferred, as a 00 has the second best fitness after a 11. Since the fitness for a 10 or 01 is the same, this does not impact the number of 1s at position i in expectation. However, if more 00s than 11s are sampled for block i, it can happen that the frequencies of block i are decreased. Since we assume that $\mu = \Omega(n \log n)$, the frequency is sufficiently high before the update and the frequencies of block i do not decrease by much with high probability (Lemma 8). Since, in the next iteration, block i is the critical block, it is then optimized within a single iteration (Lemma 9), and we do not need to worry about its frequencies decreasing again.

Neutral Frequencies. We now prove that the frequencies right of the selection-relevant block do not decrease by too much within the first n iterations.

Lemma 6. *Let $\varepsilon \in (0, 1)$ be a constant. Consider the UMDA with $\lambda \geq \mu \geq (16n/\varepsilon^2) \log n$ optimizing DECEIVINGLEADINGBLOCKS. Let $t \leq n$ be the first iteration such that block $i \in [\frac{n}{2}]$ becomes selection-relevant for the first time. Then, with a probability of at least $1 - 2n^{-1}$, all frequencies at the positions $[2i + 1..n]$ are at least $(1 - \varepsilon)/2$ within the first t iterations.*

Proof. Let $j \in [2i + 1..n]$ denote the index of a frequency right of block i. Note that by the assumption that t is the first iteration such that block i becomes selection-relevant it follows that, for all $t' \leq t$, the frequency $p_i^{(t')}$ is neutral, as we discussed above.

Since $p_j^{(t')}$ is neutral for all $t' \leq t$, by Theorem 4 with $d = \frac{\varepsilon}{2}$, we see that the probability that p_j leaves the interval $((1 - \varepsilon)/2, (1 + \varepsilon)/2)$ within the first $t \leq n$ iterations is at most $2 \exp(-\varepsilon^2 \mu/(8t)) \leq 2 \exp(-\varepsilon^2 \mu/(8n)) \leq 2n^{-2}$, where we used our bound on μ.

Applying a union bound over all $n - 2i \leq n$ neutral frequencies yields that at least one frequency leaves the interval $((1 - \varepsilon)/2, (1 + \varepsilon)/2)$ within the first t iterations with a probability of at most $2n^{-1}$, as desired. $\square$

Update of the Selection-Relevant Block. As mentioned at the beginning of the section, while frequencies right of the selection-relevant block do not drop below $(1 - \varepsilon)/2$ with high probability (by Lemma 6), the frequencies of the selection-relevant block can drop below $(1 - \varepsilon)/2$, as the following example shows.

Example 7. Consider the UMDA with $\mu = 0.05\lambda \geq c \ln n$, for a sufficiently large constant c, optimizing DLB. Consider an iteration t and assume that block $i = \frac{n}{2} - 1 - o(n)$ is critical. Assume that the frequencies in blocks i and $i+1$ are all at $2/5$. Then the offspring population in iteration t has roughly $((2/5)^2/e)\lambda \approx 0.058\lambda > \mu$ individuals with at least $2i$ leading 1s in expectation. By Theorem 1, this also holds with high probability. Thus, the frequencies in block i are set to $1 - \frac{1}{n}$ with high probability.

The expected number of individuals with at least $2i + 2$ leading 1s is roughly $((2/5)^4/e)\lambda \approx 0.0095\lambda$, and the expected number of individuals with $2i$ leading 1s followed by a 00 is roughly $((2/5)^2 \cdot (3/5)^2/e)\lambda \approx 0.02\lambda$. In total, we expect approximately $0.0295\lambda < \mu$ individuals with $2i$ leading 1s followed by either a 11 or a 00. Again, by Theorem 1, these numbers occur with high probability. Note that this implies that block $i + 1$ is selection-relevant with high probability.

Consider block $i + 1$. For selection, we choose all 0.0295λ individuals with $2i$ leading 1s followed by either a 11 or a 00 (which are sampled with high probability). For the remaining $\mu - 0.0295\lambda = 0.0205\lambda$ selected individuals with $2i$ leading 1s, we expect half of them, that is, 0.01025λ individuals to have a 10. Thus, with high probability, the frequency $p_{2i+1}^{(t+1)}$ is set to roughly $(0.0095 + 0.01025)\lambda/\mu = 0.395$, which is less than $0.4 = p_{2i+1}^{(t)}$. Thus, this frequency decreased.

The next lemma shows that such frequencies do not drop too low, however.

Lemma 8. *Let $\varepsilon, \delta \in (0, 1)$ be constants, and let c be a sufficiently large constant. Consider the UMDA with $\lambda \geq \mu \geq c \ln n$ optimizing* DECEIVINGLEADING-BLOCKS. *Further, consider an iteration t such that block $i \in [2..\frac{n}{2}]$ is selection-relevant, and assume that its frequencies $p_{2i-1}^{(t)}$ and $p_{2i}^{(t)}$ are at least $(1 - \varepsilon)/2$ when sampling the population. Then the frequencies $p_{2i-1}^{(t+1)}$ and $p_{2i}^{(t+1)}$ are at least $(1 - \delta)(1 - \varepsilon)^2/4$ with a probability of at least $1 - 4n^{-2}$.*

Proof. Let k denote the number of individuals with a prefix of at least $2i - 2$ leading 1s. Since block i is selection-relevant, it follows that $k \geq \mu$. We consider a random variable X that follows a binomial law with k trials and with a success probability of $p_{2i-1}^{(t)}p_{2i}^{(t)} =: \widetilde{p} \geq (1 - \varepsilon)^2/4$. We now bound the probability that at least $(1 - \delta)\widetilde{p}\mu =: m$ have $2i$ leading 1s, that is, we bound $\Pr[X \geq m \mid X < \mu]$, where the condition follows from the definition of block i being selection-relevant.

Elementary calculations show that

$$\Pr[X \geq m \mid X < \mu] = 1 - \Pr[X < m \mid X < \mu] = 1 - \frac{\Pr[X < m, X < \mu]}{\Pr[X < \mu]}$$

$$= 1 - \frac{\Pr[X < m]}{\Pr[X < \mu]}. \tag{1}$$

To show a lower bound for (1), consider separately the two cases that $\mathrm{E}[X] < \mu$ and $\mathrm{E}[X] \geq \mu$.

Case 1: $\mathrm{E}[X] < \mu$. We first bound the numerator of the subtrahend in (1). Since $m/(1 - \delta) = \widetilde{p}\mu \leq \widetilde{p}k = \mathrm{E}[X]$, we have $\Pr[X < m] \leq \Pr\left[X < (1 - \delta)\mathrm{E}[X]\right]$. By Theorem 1, by $\mathrm{E}[X] \geq \widetilde{p}\mu$, and by our assumption that $\mu \geq c \ln n$, choosing c sufficiently large, we have

$$\Pr\left[X < (1 - \delta)\mathrm{E}[X]\right] \leq \exp\left(-\frac{\delta^2 \mathrm{E}[X]}{2}\right) \leq \exp\left(-\frac{\delta^2 \widetilde{p}\mu}{2}\right) \leq n^{-2}.$$

For bounding the denominator, we note that $\widetilde{p} \leq 1 - \frac{1}{n}$ and use the fact that a binomially distributed random variable with a success probability of at most $1 - \frac{1}{n}$ is below its expectation with a probability of at least $\frac{1}{4}$ [7, Lemma 10.20 (b)]. This yields

$$\Pr[X < \mu] \geq \Pr\left[X < \mathrm{E}[X]\right] \geq \frac{1}{4}.$$

Combining these bounds, we obtain $\Pr[X \geq m \mid X < \mu] \geq 1 - 4n^{-2}$ for this case.

Case 2: $\mathrm{E}[X] \geq \mu > m$. We bound the subtrahend from (1) from above. By basic estimations and by Corollary 3, we see that

$$\frac{\Pr[X < m]}{\Pr[X < \mu]} \leq \frac{\Pr[X \leq m - 1]}{\Pr[X = \mu - 1]} \leq \frac{(k - m + 1)\widetilde{p}}{\mathrm{E}[X] - m + 1} \cdot \frac{\Pr[X = m - 1]}{\Pr[X = \mu - 1]}. \tag{2}$$

We bound the first factor of (2) as follows:

$$\frac{(k - m + 1)\widetilde{p}}{\mathrm{E}[X] - m + 1} \leq \frac{\mathrm{E}[X]}{\mathrm{E}[X] - m} = 1 + \frac{m}{\mathrm{E}[X] - m} \leq 1 + \frac{m}{\mu - m} \leq 1 + \frac{m}{\frac{m}{p} - m}$$

$$= 1 + \frac{\widetilde{p}}{1 - \widetilde{p}} \leq 1 + n - 1 = n,$$

where the last inequality uses that $\widetilde{p} \leq (1 - \frac{1}{n})^2 \leq 1 - \frac{1}{n}$.

For the second factor of (2), we compute

$$\frac{\Pr[X = m - 1]}{\Pr[X = \mu - 1]} = \frac{\binom{k}{m-1}\widetilde{p}^{m-1}(1 - \widetilde{p})^{k-m+1}}{\binom{k}{\mu-1}\widetilde{p}^{\mu-1}(1 - \widetilde{p})^{k-\mu+1}}$$

$$= \frac{(\mu - 1)!(k - \mu + 1)!}{(m - 1)!(k - m + 1)!} \cdot \left(\frac{1 - \widetilde{p}}{\widetilde{p}}\right)^{\mu-m}. \tag{3}$$

Since $\widetilde{p} \geq (1-\varepsilon)^2/4$, we see that $(1-\widetilde{p})/\widetilde{p} \leq 4/(1-\varepsilon)^2$.

For the first factor of (3), let $p^* := (1-\delta)\widetilde{p}$, thus $\mu p^* = m$. Noting that, for all $a, b \in \mathbb{R}$ with $a < b$, the function $j \mapsto (a+j)(b-j)$ is maximal for $j = (b-a)/2$, we first bound

$$\frac{(\mu-1)!}{(m-1)!} = \prod_{j=0}^{\mu-m-1} (\mu-1-j) \leq \prod_{j=0}^{\lfloor(\mu-m-1)/2\rfloor} ((\mu-1-j)(m+j))$$

$$\leq \left(\frac{\mu+m}{2}\right)^{\mu-m} \leq \left(\frac{\mu}{2}(1+p^*)\right)^{\mu-m}.$$

Substituting this into the first factor of (3), we bound

$$\frac{(\mu-1)!(k-\mu+1)!}{(m-1)!(k-m+1)!} \leq \left(\frac{\mu}{2}(1+p^*)\right)^{\mu-m} \cdot \frac{(k-\mu+1)!}{(k-m+1)!}$$

$$= \frac{\left(\frac{\mu}{2}(1+p^*)\right)^{\mu-m}}{\prod_{j=0}^{\mu-m-1}(k-m+1-j)} = \prod_{j=0}^{\mu-m-1} \frac{\mu(1+p^*)}{2(k-m+1-j)}.$$

By noting that $k\widetilde{p} = \mathrm{E}[X] \geq \mu$, we bound the above estimate further:

$$\prod_{j=0}^{\mu-m-1} \frac{\mu(1+p^*)}{2(k-m+1-j)} \leq \prod_{j=0}^{\mu-m-1} \frac{\mu(1+p^*)}{2(\mu/\widetilde{p}-m+1-j)}$$

$$\leq \left(\frac{\mu(1+p^*)}{2\mu(1/\widetilde{p}-1)+2}\right)^{\mu-m} \leq \left(\frac{\widetilde{p}(1+p^*)}{2(1-\widetilde{p})}\right)^{\mu-m}.$$

Substituting both bounds into (3) and recalling that $m = \mu p^*$, we obtain

$$\frac{\Pr[X = m-1]}{\Pr[X = \mu-1]} \leq \left(\frac{1+p^*}{2}\right)^{\mu(1-p^*)} = \exp\left(-\mu(1-p^*)\ln\left(\frac{2}{1+p^*}\right)\right).$$

Finally, substituting this back into our bound of (2), using our assumption that $\mu \geq c\ln n$ and noting that p^* is constant, choosing c sufficiently large, we obtain

$$\frac{\Pr[X < m]}{\Pr[X < \mu]} \leq n \exp\left(-\mu(1-p^*)\ln\left(\frac{2}{1+p^*}\right)\right) \leq n^{-2}.$$

Concluding the Proof. In both cases, we see that the number of 11s in block i is at least $m = (1-\delta)\widetilde{p}\mu \geq ((1-\delta)(1-\varepsilon)^2/4)\mu$ with a probability of at least $1 - 4n^{-2}$. Since each 11 contributes to the new values of p_{2i-1} and p_{2i}, after the update, both frequencies are at least $(1-\delta)(1-\varepsilon)^2/4$, as we claimed. □

Optimizing the Critical Block. Our next lemma considers the critical block $i \in [\frac{n}{2}]$ of an iteration t. It shows that, with high probability, for all $j \in [2i]$, we have that $p_j^{(t+1)} = 1 - \frac{1}{n}$. Informally, this means that (i) all frequencies left of the critical block remain at $1 - \frac{1}{n}$, and (ii) the frequencies of the critical block are increased to $1 - \frac{1}{n}$.

Lemma 9. *Let $\delta, \varepsilon, \zeta \in (0,1)$ be constants and let $q = (1-\delta)^2(1-\varepsilon)^4/16$. Consider the UMDA optimizing* DECEIVINGLEADINGBLOCKS *with $\lambda \geq (4/\zeta^2) \ln n$ and $\mu/\lambda \leq (1-\zeta)q/e$, and consider an iteration t such that block $i \in [\frac{n}{2}]$ is critical and that $p_{2i-1}^{(t)}$ and $p_{2i}^{(t)}$ are at least $\sqrt{q}$. Then, with a probability of at least $1 - n^{-2}$, at least μ offspring are generated with at least $2i$ leading 1s. In other words, the selection-relevant block of iteration t is at a position in $[i+1..\frac{n}{2}]$.*

Proof. Let X denote the number of individuals that have at least $2i$ leading 1s. Since block i is critical, each frequency at a position $j \in [2i-2]$ is at $1-\frac{1}{n}$. Thus, the probability that all of these frequencies sample a 1 for a single individual is $(1-\frac{1}{n})^{2i-2} \geq (1-\frac{1}{n})^{n-1} \geq 1/e$. Further, since the frequencies $p_{2i-1}^{(t)}$ and $p_{2i}^{(t)}$ are at least $\sqrt{q}$, the probability to sample a 11 at these positions is at least q. Hence, we have $\mathrm{E}[X] \geq q\lambda/e$.

We now apply Theorem 1 to show that it is unlikely that fewer than μ individuals from the current iteration have fewer than $2i$ leading 1s. Using our bounds on μ and λ, we compute

$$\Pr[X < \mu] \leq \Pr\left[X \leq (1-\zeta)\frac{q}{e}\lambda\right] \leq \Pr\left[X \leq (1-\zeta)\mathrm{E}[X]\right] \leq e^{-\frac{\zeta^2\lambda}{2}} \leq n^{-2}.$$

Thus, with a probability of at least $1 - n^{-2}$, at least μ individuals have at least $2i$ leading 1s. This concludes the proof. ∎

The Run Time of the UMDA on DLB. We now prove our main result.

Proof (of Theorem 5*).* We prove that the UMDA samples the optimum after $\frac{n}{2} + 2e \ln n$ iterations with a probability of at least $1 - 9n^{-1}$. Since it samples λ individuals each iteration, the theorem follows.

Due to Lemma 6 and $\mu \geq c_\mu n \log n$, for an $\varepsilon \in (0,1)$, within the first n iterations, with a probability of at least $1 - 2n^{-1}$, no frequency drops below $(1-\varepsilon)/2$ while its block has not been selection-relevant yet.

By Lemma 8, for another constant $\delta \in (0,1)$, with a probability of at least $1 - 4n^{-2}$, once a block becomes selection-relevant, its frequencies do not drop below $(1-\delta)(1-\varepsilon)^2/4$ for the next iteration. By a union bound, this does not fail for n consecutive times with a probability of at least $1 - 4n^{-1}$. Note that a selection-relevant block becomes critical in the next iteration.

Consider a critical block $i \in [\frac{n}{2}]$. By Lemma 9, choosing c_λ sufficiently large, with a probability of at least $1 - n^{-2}$, all frequencies at positions in $[2i]$ are immediately set to $1 - \frac{1}{n}$ in the next iteration, and the selection-relevant block has an index of at least $i+1$, thus, moving to the right. Applying a union bound for the first n iterations of the UMDA and noting that each frequency belongs to a selection-relevant block at most once shows that all frequencies are at $1 - \frac{1}{n}$ after the first $\frac{n}{2}$ iterations, since each block contains two frequencies, and stay there for at least $\frac{n}{2}$ additional iterations with a probability of at least $1 - 2n^{-1}$.

Consequently, after the first $\frac{n}{2}$ iterations, the optimum is sampled in each iteration with a probability of $(1-\frac{1}{n})^n \geq 1/(2e)$. Thus, after $2e \ln n$ additional iterations, the optimum is sampled with a probability of at least $1 - (1 - 1/(2e))^{2e \ln n} \geq 1 - n^{-1}$.

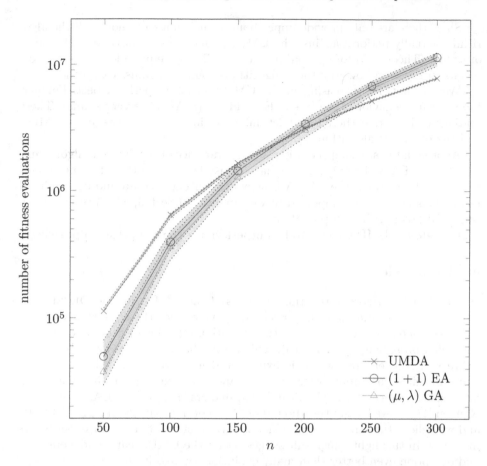

Fig. 1. This figure shows the number of fitness evaluations (note the logarithmic scale) until the optimum is sampled for the first time. Depicted are three different algorithms and for various values of n (from 50 to 300 in steps of 50), optimizing DLB. For each value of n, 50 independent runs were started per algorithm. The results of these runs are depicted above. The lines depict the median of the 50 runs of an algorithm, and the shaded areas denote the center 50%. The UMDA uses $\mu = 3n \ln n$ and $\lambda = 12\mu$. The (μ, λ) GA uses $\mu = \ln n$, $\lambda = 9\mu$, uniform crossover, and has a crossover probability of $1/2$.

Overall, by applying a union bound over all failure probabilities above, the UMDA needs at most $n + \ln n$ iterations to sample the optimum with a probability of at least $1 - 9n^{-1}$. □

4   Experiments

In their paper, Lehre and Nguyen [18] analyze (among others) the $(1 + 1)$ EA and the (μ, λ) GA on DLB. For an optimal choice of parameters, they prove an expected run time of $O(n^3)$ for all considered algorithms.

Since these are only provide upper bounds, it is not clear how well the algorithms actually perform against the UMDA, which has a run time in the order of $n^2 \ln n$ (Theorem 5) for optimal parameters. Thus, we provide some empirical results in Fig. 1 on how well these algorithms compare against each other.

We see that, for increasing n, the UMDA seems to perform best. Further, the run time behavior of the $(1+1)$ EA and the (μ, λ) GA is very similar. These findings indicate that there is a strict difference in run time between the UMDA and the other two algorithms.

Another interesting aspect of Fig. 1 is the variance of the different algorithms. The $(1+1)$ EA and the (μ, λ) GA have a visible variance that does not seems to reduce. In contrast, the UMDA is strongly concentrated around its empirical mean. This behavior is supported by our analysis in Sect. 3, which proves a run time that holds with high probability.

Overall, the UMDA seems to be outperforming the competing approaches.

5   Conclusion

We conducted a rigorous run time analysis of the UMDA on the DECEIVING-LEADINGBLOCKS function. In particular, it shows that the algorithm with the right parameter choice finds the optimum within $O(n^2 \log n)$ fitness evaluations with high probability (Theorem 5). This result shows that the lower bound by Lehre and Nguyen [18], which is exponential in μ, is not due to the UMDA being ill-suited for coping with epistasis and deception, but rather due to an unfortunate choice of the algorithm's parameters. For several EAs, Lehre and Nguyen [18] showed a run time bound of $O(n^3)$ on DECEIVINGLEADINGBLOCKS and we believe that this is tight, which is further supported by our experiments in Sect. 4. In this light, our result suggests that the UMDA can handle epistasis and deception even better than many evolutionary algorithms.

Our run time analysis holds for parameter regimes that prevent genetic drift. When comparing our run time with the one shown in [18], we obtain a strong suggestion for running EDAs in regimes of low genetic drift. In contrast to the work of Lengler, Sudholt, and Witt [19] that indicates moderate performance losses due to genetic drift, here we obtain the first fully rigorous proof of such a performance loss, and in addition one that is close to exponential in n (the $\exp(-\Omega(\mu))$ lower bound of [18] holds for μ up to $o(n)$).

On the technical side, our result indicates that the regime of low genetic drift admits relatively simple and natural analyses of run times of EDAs, in contrast, e.g., to the level-based methods previously used in comparable analyses, e.g., in [4, 18].

We conjecture that our result can be generalized to a version of the DECEI-VINGLEADINGBLOCKS function with a block size of $k \le n$.

Acknowledgments. This work was supported by COST Action CA15140 and by a public grant as part of the Investissements d'avenir project, reference ANR-11-LABX-0056-LMH, LabEx LMH, in a joint call with Gaspard Monge Program for optimization, operations research and their interactions with data sciences.

References

1. Chen, T., Lehre, P.K., Tang, K., Yao, X.: When is an estimation of distribution algorithm better than an evolutionary algorithm? In: Proceedings of CEC 2009, pp. 1470–1477 (2009). https://doi.org/10.1109/CEC.2009.4983116
2. Dang, D., et al.: Escaping local optima with diversity mechanisms and crossover. In: Proceedings of GECCO 2016, pp. 645–652 (2016). https://doi.org/10.1145/2908812.2908956
3. Dang, D., et al.: Escaping local optima using crossover with emergent diversity. IEEE Trans. Evol. Comput. **22**(3), 484–497 (2018). https://doi.org/10.1109/TEVC.2017.2724201
4. Dang, D., Lehre, P.K.: Simplified runtime analysis of estimation of distribution algorithms. In: Proceedings of GECCO 2015, pp. 513–518 (2015). https://doi.org/10.1145/2739480.2754814
5. Doerr, B.: Analyzing randomized search heuristics via stochastic domination. Theor. Comput. Sci. **773**, 115–137 (2019). https://doi.org/10.1016/j.tcs.2018.09.024
6. Doerr, B.: A tight runtime analysis for the cGA on jump functions: EDAs can cross fitness valleys at no extra cost. In: Proceedings of GECCO 2019, pp. 1488–1496 (2019). https://doi.org/10.1145/3321707.3321747
7. Doerr, B.: Probabilistic tools for the analysis of randomized optimization heuristics. In: Doerr, B., Neumann, F. (eds.) Theory of Evolutionary Computation. NCS, pp. 1–87. Springer, Cham (2020). https://doi.org/10.1007/978-3-030-29414-4_1. https://arxiv.org/abs/1801.06733
8. Doerr, B., Krejca, M.S.: Significance-based estimation-of-distribution algorithms. In: Proceedings of GECCO 2018, pp. 1483–1490 (2018). https://doi.org/10.1145/3205455.3205553
9. Doerr, B., Künnemann, M.: Optimizing linear functions with the $(1+\lambda)$ evolutionary algorithm - different asymptotic runtimes for different instances. Theor. Comput. Sci. **561**, 3–23 (2015). https://doi.org/10.1016/j.tcs.2014.03.015
10. Doerr, B., Zheng, W.: Sharp bounds for genetic drift in EDAs. CoRR abs/1910.14389 (2019). https://arxiv.org/abs/1910.14389
11. Droste, S.: A rigorous analysis of the compact genetic algorithm for linear functions. Nat. Comput. **5**(3), 257–283 (2006). https://doi.org/10.1007/s11047-006-9001-0
12. Droste, S., Jansen, T., Wegener, I.: On the analysis of the $(1+1)$ evolutionary algorithm. Theor. Comput. Sci. **276**(1–2), 51–81 (2002). https://doi.org/10.1016/S0304-3975(01)00182-7
13. Hasenöhrl, V., Sutton, A.M.: On the runtime dynamics of the compact genetic algorithm on jump functions. In: Proceedings of GECCO 2018, pp. 967–974 (2018). https://doi.org/10.1145/3205455.3205608
14. Hoeffding, W.: Probability inequalities for sums of bounded random variables. J. Am. Stat. Assoc. **58**(301), 13–30 (1963). https://doi.org/10.2307/2282952
15. Krejca, M.S., Witt, C.: Lower bounds on the run time of the univariate marginal distribution algorithm on OneMax. In: Proceedings of FOGA 2017, pp. 65–79 (2017). https://doi.org/10.1145/3040718.3040724
16. Krejca, M.S., Witt, C.: Theory of estimation-of-distribution algorithms. In: Doerr, B., Neumann, F. (eds.) Theory of Evolutionary Computation. NCS, pp. 405–442. Springer, Cham (2020). https://doi.org/10.1007/978-3-030-29414-4_9
17. Lehre, P.K., Nguyen, P.T.H.: Improved runtime bounds for the univariate marginal distribution algorithm via anti-concentration. In: Proceedings of GECCO 2017, pp. 1383–1390 (2017). https://doi.org/10.1145/3071178.3071317

18. Lehre, P.K., Nguyen, P.T.H.: On the limitations of the univariate marginal distribution algorithm to deception and where bivariate EDAs might help. In: Proceedings of FOGA 2019, pp. 154–168 (2019). https://doi.org/10.1145/3299904.3340316

19. Lengler, J., Sudholt, D., Witt, C.: Medium step sizes are harmful for the compact genetic algorithm. In: Proceedings of GECCO 2018, pp. 1499–1506 (2018). https://doi.org/10.1145/3205455.3205576

20. Mühlenbein, H., Paaß, G.: From recombination of genes to the estimation of distributions I. Binary parameters. In: Proceedings of PPSN 1996, pp. 178–187 (1996). https://doi.org/10.1007/3-540-61723-X_982

21. Pelikan, M., Hauschild, M.W., Lobo, F.G.: Estimation of distribution algorithms. In: Kacprzyk, J., Pedrycz, W. (eds.) Springer Handbook of Computational Intelligence, pp. 899–928. Springer, Heidelberg (2015). https://doi.org/10.1007/978-3-662-43505-2_45

22. Sudholt, D., Witt, C.: On the choice of the update strength in estimation-of-distribution algorithms and ant colony optimization. Algorithmica **81**(4), 1450–1489 (2018). https://doi.org/10.1007/s00453-018-0480-z

23. Witt, C.: Domino convergence: why one should hill-climb on linear functions. In: Proceedings of GECCO 2018, pp. 1539–1546 (2018). https://doi.org/10.1145/3205455.3205581

24. Witt, C.: Upper bounds on the running time of the univariate marginal distribution algorithm on OneMax. Algorithmica **81**(2), 632–667 (2018). https://doi.org/10.1007/s00453-018-0463-0

25. Zheng, W., Yang, G., Doerr, B.: Working principles of binary differential evolution. In: Proceedings of GECCO 2018, pp. 1103–1110 (2018). https://doi.org/10.1145/3205455.3205623

A Beam Search Approach
to the Traveling Tournament Problem

Nikolaus Frohner$^{(\boxtimes)}$, Bernhard Neumann, and Günther R. Raidl

Institute of Logic and Computation, TU Wien, Vienna, Austria
{nfrohner,raidl}@ac.tuwien.ac.at, e1634034@student.tuwien.ac.at

Abstract. The well-known traveling tournament problem is a hard optimization problem in which a double round robin sports league schedule has to be constructed while minimizing the total travel distance over all teams. The teams start and end their tours at their home venues, are only allowed to play a certain maximum number of games in a row at home or away, and must not play against each other in two consecutive rounds. The latter aspects introduce also a difficult feasibility aspect. In this work, we study a beam search approach based on a recursive state space formulation. We compare different state ordering heuristics for the beam search based on lower bounds derived by means of decision diagrams. Furthermore, we introduce a randomized beam search variant that adds Gaussian noise to the heuristic value of a node for diversifying the search in order to enable a simple yet effective parallelization. In our computational study, we use randomly generated instances to compare and tune algorithmic parameters and present final results on the classical National League and circular benchmark instances. Results show that this purely construction-based method provides mostly better solutions than existing ant-colony optimization and tabu search algorithms and it comes close to the leading simulated annealing based approaches without using any local search. For two circular benchmark instances we found new best solutions for which the last improvement was twelve years ago. The presented state space formulation and lower bound techniques could also be beneficial for exact methods like A* or DFS* and may be used to guide the randomized construction in ACO or GRASP approaches.

Keywords: Traveling tournament problem · Beam search · Decision diagrams

1 Introduction

In 2001, Easton, Nemhauser, and Trick [4] introduced the traveling tournament problem (TTP). It concerns the construction of a double round robin schedule for a sports league, where the sum of the travel distances over all teams shall be minimized. Teams start and end at their respective home venues and are assumed to always travel directly from their current position to their next designated game venue, which is either at home or away. They are only allowed to play a certain

© Springer Nature Switzerland AG 2020
L. Paquete and C. Zarges (Eds.): EvoCOP 2020, LNCS 12102, pp. 67–82, 2020.
https://doi.org/10.1007/978-3-030-43680-3_5

maximum number of games away or at home consecutively, and two teams must not play against each other in two subsequent rounds. These aspects make even finding any feasible schedule in general difficult. At the time of writing, proven optimal solutions have been found for classical benchmarks instances with up to ten teams, but not for twelve and more teams, as stated on Michael Trick's TTP web page[1].

Due to the problem's complexity, many different metaheuristics have already been suggested to solve larger instances approximately. Neighborhood search based approaches as tabu search [3] or simulated annealing [1,14] provide particularly strong results. In this contribution, we present a beam search based on a new recursive state space formulation of the problem. We compare different lower bound heuristics to order the nodes in a layer of the state graph, which is traversed in breadth-first-search manner. Competitive results can be achieved with a randomized variant of the beam search in which we add noise to the heuristic estimates. This randomization enables a simple yet effective execution of multiple diversified beam search runs in parallel.

In Sect. 2 we summarize the work on which our new approach is based, specifically worth mentioning are the papers of Uthus, Riddle, and Guesgen [11–13], which broadly fall into the class of tree search based techniques. In particular, we build upon their bound pre-calculation method. We formally introduce the TTP in Sect. 3 and give an associated state space formulation in Sect. 4, which differs from integer programming or constraint programming formulations primarily used so far. A state may be reached by different partial schedules and determines the feasible completions to a complete schedule, which allows to detect and break symmetries on the go. Section 5 is concerned with the schedule construction algorithm on the state graph using beam search driven by lower bounds that are derived from the states. These lower bounds are calculated by solving either an associated traveling salesperson problem (TSP) or a capacitated vehicle routing problem (CVRP) independently for each team; the latter corresponds to the well-known independent lower bound (ILB) introduced by Easton et al. [4].

We introduce a method to pre-calculate lower bounds for all states by means of decision diagrams. Moreover, we show in Sect. 6 how these bounds can be further tightened using the minimum number of trips (MNT) bound introduced by Urrutia et al. [10]. Section 7 presents computational results. We tune algorithmic parameters and compare the performance of different algorithm variants on randomly generated instances on a two-dimensional grid, and conduct final tests on the classical benchmark instances derived from teams of the US Major League Baseball (NL) and the circular instances (CIRC) [4]. We observe that our purely constructive approach, which does not make use of any local search, delivers competitive results. In particular, we could find new best solutions for two circular instances. Finally, we conclude in Sect. 8 and make suggestions for further research.

---

[1] https://mat.tepper.cmu.edu/TOURN/.

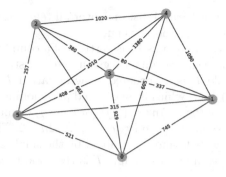

$$\begin{pmatrix} 5 & -3 & 2 & 6 & -1 & -4 \\ -3 & 6 & 1 & 5 & -4 & -2 \\ -2 & 1 & -4 & 3 & -6 & 5 \\ -5 & 3 & -2 & -6 & 1 & 4 \\ 4 & -5 & 6 & -1 & 2 & -3 \\ 2 & -1 & 4 & -3 & 6 & -5 \\ 6 & -4 & 5 & 2 & -3 & -1 \\ -4 & 5 & -6 & 1 & -2 & 3 \\ -6 & 4 & -5 & -2 & 3 & 1 \\ 3 & -6 & -1 & -5 & 4 & 2 \end{pmatrix}$$

Fig. 1. Left: the NL6 problem instance from [4] shown as complete undirected graph. Right: a feasible double round robin tournament schedule represented by a $(2n-2) \times n$ matrix, where the value j of entry (r, i) corresponds to the game $i \rightarrow^r |j|$, if j is negative, otherwise to the game $j \rightarrow^r i$.

2   Previous Work

The TTP itself, together with the NL and CIRC benchmark instances, and the ILB were introduced by Easton et al. [4]. The MNT lower bound was proposed by Urrutia et al. [10] including an algorithm to calculate it. Uthus et al. [13] suggested an exact iterative deepening A* search, which allowed them to solve the NL instance with ten teams to proven optimality. Their approach features special symmetry breaking techniques, memoization, and was performed in parallel on 120 processors for a wall time of roughly 67 h.

From this work, we adopt the method to pre-calculate independent lower bounds for states to occur during the state space traversal. We aim at solving larger instances approximately and compare our beam search results therefore to the results of today's state-of-the-art metaheuristic approaches, which are the simulated annealing from [1], the tabu search from [3], the ant colony optimization from [11], and the population-based simulated annealing from [14], where the latter found the so far best solutions for the larger NL and CIRC instances using a cluster consisting of 60 nodes.

For beam search in general, see, e.g., [6]. For a thorough introduction to decision diagrams in combinatorial optimization, we recommend the book by Bergman et al. [2].

3   Problem Formalization

We are given a set $V = \{1, \ldots, n\}$ of n teams, where n is even, and a distance matrix d where $d(i, j)$ is the traveling distance from team i's home venue to team j's home venue, $i, j \in V$. The goal is to find a double round robin tournament schedule, where every team plays at most U games subsequently at home or on the road (*at-most*), respectively, teams must not play against each other in subsequent rounds (*no-repeat*), and the total travel distance over all teams is to be minimized. Each team starts and ends at its home venue.

Adopting the formulation of [8], we see the teams V as vertices of a complete weighted directed graph $G = (V, A)$, where the weights are given by the distance matrix d. A double round robin schedule T is an ordered 1-factorization $T = (G^1 = (V, A^1), \ldots, G^{2n-2} = (V, A^{2n-2}))$ of G, which is an ordered partitioning of the arcs into $2n-2$ perfect matchings (1-factors). An arc (i, j) (or $i \rightarrow^r j$) denotes that team i plays against team j at j's venue in round r, $r = 1, \ldots, 2n - 2$. The location of team i in round r is denoted $p_i^r \in V$ and determined by the single arc in A^r incident to team i. The objective value of a schedule T is the total travel distance given by

$$z(T) = \sum_{i=1}^{n} \left(d(i, p_i^1) + \sum_{r=2}^{2n-2} d(p_i^{r-1}, p_i^r) + d(p_i^{2n-2}, i) \right). \tag{1}$$

Throughout this paper and as in most previous work, we only consider $U = 3$, for which Thielen and Westphal [9] have shown strong NP-completeness in the corresponding decision variant of the problem.

Figure 1 shows on the left an example instance with $n = 6$ teams depicted as an undirected complete graph (distances are here assumed to be symmetric). A corresponding feasible TTP schedule is shown on the right, represented as a $(2n - 2) = 10$ rounds by six teams matrix, denoting the opponent and venue for each round and team.

4   State Space Formulation

We model the solution space, i.e., the set of feasible schedules of a TTP instance (V, d), by a state graph. This is a rooted directed acyclic graph representing the feasible schedules by corresponding paths from a root state to a dedicated terminal state. The states (nodes) are organized into $n^2 - n + 2$ layers, where layer 0 only contains the root state s_r, layer $n^2 - n + 1$ only the terminal state s_t, and layers $l = 1, \ldots, n^2 - n$ contain states representing the situations after the l-th played game.

Each state is a tuple $(\mathbf{M}^s, \mathbf{y}^s, \mathbf{r}^s, \mathbf{x}^s, \mathbf{h}^s, \mathbf{o}^s)$, where $\mathbf{M}^s = (M_{i,j}^s)_{i,j \in V} \in \{0, 1\}^{n \times n}$ is an incidence matrix that indicates the games left to be scheduled and vectors $\mathbf{y}^s = (y_i^s)_{i \in V}$, $\mathbf{r}^s = (r_i^s)_{i \in V}$, $\mathbf{x}^s = (x_i^s)_{i \in V}$, $\mathbf{h}^s = (h_i^s)_{i \in V}$, and $\mathbf{o}^s = (o_i^s)_{i \in V}$ represent for each team i the currently forbidden opponent y_i^s, the current round r_i^s, its location x_i^s, and the number of still possible home or away games left to play in a row h_i^s and o_i^s, respectively. The forbidden opponents $\mathbf{y}^s$ are used to implement the *no-repeat* constraint and $\mathbf{h}^s$ and $\mathbf{o}^s$ to take care of the *at-most* constraints. Moreover, this information contained in a state implies for each team $i \in V$ the set P_i^s of the games that can be played next without violating the TTP constraints.

A state transition from a state s at layer l to a state s' at layer $l + 1$, $l = 0, \ldots, n^2 - n$, corresponds to a specific game $i \rightarrow^r j$ being played by teams

Partial schedule

$$
\begin{pmatrix}
5 & -3 & 2 & 6 & -1 & -4\\
-3 & 6 & 1 & 5 & -4 & -2\\
-2 & 1 & -4 & 3 & -6 & 5\\
- & - & - & - & - & -\\
- & - & - & - & - & -\\
- & - & - & - & - & -\\
- & - & - & - & - & -\\
- & - & - & - & - & -\\
- & - & - & - & - & -\\
- & - & - & - & - & -\\
- & - & - & - & - & -
\end{pmatrix}
\qquad
\mathbf{M}^s =
\begin{pmatrix}
0 & 0 & 0 & 1 & 1 & 1\\
1 & 0 & 0 & 1 & 1 & 1\\
1 & 1 & 0 & 0 & 1 & 1\\
1 & 1 & 1 & 0 & 1 & 1\\
0 & 1 & 1 & 0 & 0 & 1\\
1 & 0 & 1 & 0 & 1 & 0
\end{pmatrix}
\rightarrow
\mathbf{M}^{s'} =
\begin{pmatrix}
0 & 0 & 0 & 1 & 1 & 1\\
1 & 0 & 0 & 1 & 1 & 1\\
1 & 1 & 0 & 0 & 1 & 1\\
1 & 1 & 1 & 0 & 1 & 1\\
0 & 1 & 1 & 0 & 0 & 0\\
1 & 0 & 1 & 0 & 1 & 0
\end{pmatrix}
$$

$$
\mathbf{x}^s =
\begin{pmatrix}1\\1\\1\\3\\3\\1\end{pmatrix}
\rightarrow
\mathbf{x}^{s'} =
\begin{pmatrix}1\\1\\1\\3\\5\\5\end{pmatrix}
\quad
\mathbf{o}^s =
\begin{pmatrix}1\\1\\3\\3\\1\\1\end{pmatrix}
\rightarrow
\mathbf{o}^{s'} =
\begin{pmatrix}1\\1\\3\\2\\0\\3\end{pmatrix}
\quad
\mathbf{h}^s =
\begin{pmatrix}3\\1\\3\\0\\3\\3\end{pmatrix}
\rightarrow
\mathbf{h}^{s'} =
\begin{pmatrix}3\\1\\3\\0\\3\\2\end{pmatrix}
$$

Fig. 2. Left: exemplary partial schedule for an instance with six teams before ending the third round, for which the teams six and five (in bold) are selected to play the next game. Right: corresponding state updates where in the matrix of the games left the currently forbidden games implied by $\mathbf{y}^s, \mathbf{o}^s, \mathbf{h}^s$ are grayed out. We omitted $\mathbf{r}^s$ and $\mathbf{y}^s$ for space reasons.

i and j at j's venue in round r. Each state transition is weighted by the sum of the distances both teams have to travel from their previous locations to play the game

$$\Delta z(s,s') = d(x_i^s, x_i^{s'}) + d(x_j^s, x_j^{s'}). \tag{2}$$

Teams for the game are selected in a way that the partial schedule grows round by round in ascending order where each round is completed before the next one starts. All paths starting from the root state and leading to the terminal state correspond to feasible solutions. Paths that end before the terminal state at a state without further transitions represent partial schedules that cannot be feasibly continued. A shortest path from the root to the terminal state therefore corresponds to an optimal feasible solution for a given problem instance.

We introduce two special rounds $r = 0$ and $r = 2n - 1$ where every team is at its home location. Let $\mathbf{M}^{s_r}$ be the matrix with non-diagonal ones and diagonal zeros, corresponding to all games to be played, and matrix $\mathbf{M}^{s_t}$ be the all-zeros matrix. If there is no forbidden opponent for a team $i \in V$ in state s, then y_i^s is set to -1. The root state is then $s_r = (\mathbf{M}^{s_r}, \mathbf{y}^{s_r} = (-1, \ldots, -1), \mathbf{r}^{s_r} = (0, \ldots, 0), \mathbf{x}^{s_r} = (1, \ldots, n), \mathbf{h}^{s_r} = (U, \ldots, U), \mathbf{o}^{s_r} = (U, \ldots, U))$ and the terminal state $s^t = (\mathbf{M}^{s_t}, \mathbf{y}^{s_t} = (-1, \ldots, -1), \mathbf{r}^{s_t} = (2n - 1, \ldots, 2n - 1), \mathbf{x}^{s_t} = (1, \ldots, n), \mathbf{h}^{s_t} = (0, \ldots, 0), \mathbf{o}^{s_t} = (0, \ldots, 0))$. Transitions to the terminal state are special in the sense that they do not correspond to played games but just to going back to the teams' home venues.

Transitions from a state s at some layer l to a subsequent state s' at layer $l + 1$ are done by selecting a game $(i, j) \in P_i^s$ (or (j, i)) where we impose the condition $r_i = r_j = \min_{i \in V} r_i$. This ensures that the teams are in the same round and games are assigned to teams round by round. If there exists a *dead* team i with $P_i^s = \emptyset$, our current state has no feasible completion. Since there is no meaning in which order we select the teams in a specific round r, we break this symmetry by defining a specific team permutation $\pi \colon V \to V$. At each state of layer l, a game from $P_{\pi_i}^s$ has to be played for which i and r_{π_i} are minimal. A trivial ordering is the lexicographic ordering of the teams.

Selecting the game (i,j) yields state s' with $\mathbf{M}^{s'}$ being a copy of $\mathbf{M}^s$ except that $M_{i,j}^{s'} = 0$, which implicitly removes this game from $P_i^{s'}$ and $P_j^{s'}$ as well. The position, round, and streak related information of i and j is updated from s to s' accordingly. To respect the *no-repeat* constraints, the forbidden opponent vector $\mathbf{y}^s$ is copied to $\mathbf{y}^{s'}$ except that $y_i^{s'} = j$ and $y_j^{s'} = i$, if $M_{j,i}^s = 1$; otherwise these values are set to -1. For every other team $k \in V \setminus \{i,j\}$, $y_k^{s'} = -1$ is set if $y_k^s \in \{i,j\}$. The *at-most* constraints are already implied by the updates in $\mathbf{o}^{s'}$ and $\mathbf{h}^{s'}$. If $o_i^{s'} = 0$, then away games are not allowed in the next round for team i; analogously, a continuation of j's home stand is not allowed, if $h_j^{s'} = 0$.

An exemplary state transition is shown in Fig. 2 for an instance with six teams before and after ending the third round with the game $(5,6)$. We see that team five hits its away streak limit and all its away games are not available for the next round and that the game $(6,5)$ is forbidden.

5   Beam Search

We perform a layer-by-layer breadth-first-search traversal of the state graph, where for each state all permitted games for a selected team are played by performing the respective transitions to corresponding successor states. The current shortest path value and the corresponding partial schedule are cached for each state during construction and updated if a shorter path to an already visited state is discovered.

Due to the complexity of the problem, only instances with four teams admit a complete construction of the state graph, providing a guaranteed optimal solution. We therefore restrict the search to an incomplete beam search where at each layer at most β states are kept for further consideration; parameter β is hereby called the beam width. In this way the total number of expanded states is polynomially bounded by $\mathcal{O}(n^2\beta)$. The shortest path through such a restricted state graph then corresponds to a feasible heuristic solution. To guide the search, in each layer the β most promising states are kept according to some state ranking heuristic, in the hope that the finally shortest path corresponds to an optimal or close-to-optimal solution. Classical beam search sorts the states by an f-value known from A* search that combines the length of the currently shortest path $g(s)$ to the state s with a lower bound $b(s)$ (or heuristic estimate) for the further continuation to the terminal state:

$$f(s) = g(s) + b(s) \tag{3}$$

In our beam search implementation, we only keep the current layer in a queue and the successive layer in a priority queue sorted according to f that contains at most the β best successor states so far. The latter is implemented by a maximum heap combined with a hash map to access arbitrary states in expected constant time. Before creating a successor state, we check in case of a full beam by means of incremental evaluation whether the potential new state's f-value is worse than the worst f-value in the heap. If this is the case, we do not need to consider the state further. Otherwise we create the successor state and check whether it already exists in the maximum heap, in which case we conditionally update its shortest path value and current best partial schedule. If the state was not yet contained in the heap, we replace its so far worst state by the newly created successor state. This approach gives us a smaller memory footprint than storing all created states until termination, allowing us to test higher beam widths. The current partial schedule is cached along each state in a growing vector.

As will be discussed in detail in Sect. 6, the lower bound values are also cached along the state for each team, together with the number of home and away games left for each team. The latter allow to quickly check whether there are not enough home games in relation to away games or vice versa to make a feasible completion.

To allow a simple parallelization of the beam search by independent diversified runs, we further introduce a randomized variant of the beam search. To this end we add a normally distributed random offset with standard deviation σ to each state's original f-value:

$$\tilde{f}(s) = f(s) + \mathcal{N}(0, \sigma). \tag{4}$$

The motivation is that states which would be pruned when just considering their deterministic f-value get a chance to survive, and they may possibly lead to superior solutions. Initially promising states can also get cut off early by drawing a too high random offset. Crucial is the standard deviation σ for which we make the following parameterized ansatz:

$$\sigma = \sigma_{\text{rel}} \cdot b(s^{\text{r}}) \tag{5}$$

Parameter σ_{rel} thus determines the fraction of the lower bound of the root state to be used as σ, so that the order of magnitude of the expected solution length of a given instance is respected. Tuning results for this parameter are presented in the computational study in Sect. 7.

Algorithm 1 shows our beam search in pseudo-code. next-team(l, s) selects the team to consider for a given layer l and state s. Trivial options are to take the lexicographically smallest team that is in a minimal round or to initially fix a random permutation of the teams.

Procedure feasibility-and-optimality-check$(H, \beta, s, b, \epsilon, (i, j))$ incrementally checks whether the transition would lead to a state for which we know for sure that it does not have a feasible completion. This is the case when not enough home or away games are available for a specific team to not violate the at-most constraint or because one team has an empty possible games set in this round. The optimality check is done by considering the increase in the f-value by the move and whether it is worse than the maximal f-value in a full, i.e., containing

Input: number of teams n, distance matrix d, root state s^{r}, terminal state s^{t},
          noise parameter σ_{rel}, state lower bound function b, beam width β
Output: feasible schedule T

```
 1 queue Q ← {sʳ};
 2 for l ← 1 to n² − n do
 3 │   H ← empty maximum heap;
 4 │   while Q ≠ ∅ do
 5 │   │   s ← Q.pop;
 6 │   │   t ← next-team(l, s);
 7 │   │   foreach (i, j) ∈ {(i′, j′) ∈ Pₜˢ|rᵢ′ˢ = rⱼ′ˢ} do
 8 │   │   │   ε ← 𝒩(0, σ_rel · b(sʳ));
 9 │   │   │   if feasibility-and-optimality-check (H, β, s, b, ε, (i, j)) then
10 │   │   │   │   s′ ← copy s and make transition by playing (i, j) and updating
                       state along with cached data accordingly;
11 │   │   │   │   s′.current_schedule ← s′.current_schedule ∪ (i, j);
12 │   │   │   │   f(s′) ← g(s′) + b(s′) + ε;
13 │   │   │   │   include s′ into H respecting f(s′);
14 │   │   │   │   if H.size > β then
15 │   │   │   │   │   remove worst element of H;
16 │   │   │   │   end
17 │   │   │   end
18 │   │   end
19 │   end
20 │   Q ←sorted-by-f-value(H);
21 end
22 if Q ≠ ∅ then
23 │   create going home transitions for all states Q to sᵗ;
24 │   return sᵗ.current_schedule;
25 else
26 │   return ∅;
```

Algorithm 1. Beam search for the TTP.

β states, maximum heap H—then the transition for game (i, j) does not need to be considered and state s' is not created, which saves a costly state expansion operation that would require to copy the whole current state and cache variables.

After all successor states have been checked and potentially included into the heap H, its states are transferred to queue Q, sorted according to state priorities, and thus these nodes become the new current layer. The sorting is done to fill the beam earlier with likely better states to increase the odds for rejecting the creation of successor states during incremental evaluation.

In the next section, we study in detail the crucial part of devising and efficiently calculating a lower bound b for determining the state's f-value.

6   Lower Bounds Calculation

The main idea for obtaining lower bounds is to relax the problem by considering the tours of all teams independently. Easton et al. [4] already suggested the independent lower bound (ILB) that applies this principle. This bound neglects the *no-repeat* constraints and considers only the away teams for a given team $i \in V$ with only the away *at-most* constraints. This amounts to a capacitated vehicle routing problem (CVRP), where the depot is at i's home venue, the customers are the away teams with unit demand, and the capacity for the trucks is $U = 3$. The CVRP itself is strongly NP-hard but for few customers tractable in practice.

Given an arbitrary state s and team i, we have to consider the remaining away teams $\mathcal{A}_i^s$ for i, the position x_i^s, and the remaining away streak o_i^s. If $x_i^s \neq i \wedge o_i^s = 0$, then we consider an artificial state in which the team is assumed to have returned home (this is the only option it has at that moment), $o_i^s = \min(\mathcal{A}_i^s, U)$, and add $d(x_i^s, i)$ to the resulting bound. Let the optimal total length for this problem for team i be $b_i^{\mathrm{CVRP}}(s)$. Then the sum of the optimal values over all teams is a lower bound for the optimal value of the corresponding TTP-feasible completion of s

$$b^{\mathrm{CVRP}}(s) = \sum_{i=1}^{n} b_i^{\mathrm{CVRP}}(s). \tag{6}$$

A natural further relaxation is to drop even the away *at-most* constraints, which yields a traveling salesperson problem based lower bound b^{TSP}. In this case, we do not have to consider the current away streak of team i in state s.

To provide better guidance for the beam search, we are more interested in tighter lower bounds, while keeping their computational costs in mind. A first natural strengthening is to consider also the home *at-most* constraints. Let $h_i^{\mathrm{left}} = |\mathcal{H}_i^s|$ be the number of home games left for team i in state s. Then we need at least $\tilde{h}_i^{\min} = \lceil (h_{\mathrm{left}} + \bar{h}_i)/U \rceil$ home stands to accommodate for the home games, where $\bar{h}_i$ is the length of the current home stand. Translated to the CVRP, this amounts to the constraint that we need to perform at least $\tilde{h}$ non-trivial tours. Analogously, every away streak needs at least one home game from where it came, minus one if the team is currently at home. This gives us a maximum to the home stands $\tilde{h}_i^{\max}$ we can realize from a given state. We can therefore define a home stand constrained lower bound $b_i^{\mathrm{CVRPC}}(s, \tilde{h})$ and tighten the CVRP bound by finding the minimum within the range of allowed home stands, summed over all teams resulting in the CVRP with home stands bound (CVRPH)

$$b^{\mathrm{CVRPH}}(s) = \sum_{i=1}^{n} \min_{\tilde{h} \in \{\tilde{h}_i^{\min}, \dots, \tilde{h}_i^{\max}\}} b_i^{\mathrm{CVRPC}}(s, \tilde{h}). \tag{7}$$

To speed up our beam search, we pre-calculate the lower bounds for the states that can occur for a given TTP instance, similarly as done by Uthus et al. [13].

We do this by representing the whole space of feasible solutions to the given CVRP instance for each team i with an exact multi-valued decision diagram (DD) [2] and finally store the lower bounds for the states that occurred in a lookup table. Each node in this DD is associated with a state q consisting of the away games left to play $\mathcal{A}^q$ (represented by the subset of other teams against which team i still has to play), the team i's position x^q and the current number of consecutive away games, the away streak $\bar{o}^q$. The root state for a given team i is therefore $q_r = (\{1, \ldots, i-1, i+1, \ldots, n\}, i, 0)$. Transitions are made until the terminal state $q_t = (\{\}, i, 0)$ is reached, where in every layer, all available transitions are performed. Hereby we distinguish between three possibilities:

- Select any away team left $j \in \mathcal{A}^q$ to visit next, if there are such, and go home afterwards, where costs $d(x^q, j) + d(j, i)$ accrue. The state is updated accordingly to $(\mathcal{A}^q \setminus \{j\}, i, 0)$.
- If $\bar{o}^q$ is less than $U - 1$, then select any away team left $j \in \mathcal{A}^q$ to visit next, if there are such, and stay at j afterwards, where costs $d(x^q, j)$ accrue. The state is updated to $(\mathcal{A}^q \setminus \{j\}, j, \bar{o}^q + 1)$.
- If $\mathcal{A}^q$ is empty, go home if not already at home, where costs $d(x^q, i)$ accrue. The state is updated to the terminal state q_t.

Paths from the root to the terminal node in the DD then correspond to the feasible solutions of the CVRP, and with the costs associated with the transitions (i.e., arcs in the DD), the lengths of such paths correspond to solution lengths. For each node in the CVRP decision diagram the shortest path to the terminal node is calculated and saved in the lookup table, serving as a lower bound for a team with given away teams to play, being at a position either at home or at some away team and its current away streak. Being a layered directed acyclic multigraph, the shortest paths for each node in the decision diagram can be calculated efficiently by doing a breadth-first-search backwards from the terminal to the root node.

The TSP based bound values can also be pre-calculated by these method by simply ignoring the away streak and allowing always a direct transition to a next away team without going home first.

Furthermore, for the CVRPH bound, we consider constrained shortest path lengths $z^{\text{sp}}(q, \tilde{h})$ from any node to the terminal node, with the constraint that exactly $\tilde{h}$ home stands occur. This means that at most $\tilde{h}U - \bar{h}_i$ home games can be played from a given node, where $\bar{h}_i$ is the length of the current home stand for team i. At the terminal node $z^{\text{sp}}(q_t, 1) = 0$ and ∞ for each other node. In the backward sweep from q' to q with arc costs $c_{q,q'}$, if $x_q \neq i$, then we set $z^{\text{sp}}(q, \tilde{h}) = \min\{z^{\text{sp}}(q', \tilde{h}) + c_{q,q'}, z^{\text{sp}}(q, \tilde{h})\}$. If on the other hand $x^q = i$, i.e., a new home stand has occurred, we set $z^{\text{sp}}(q, \tilde{h} + 1) = \min\{z^{\text{sp}}(q', \tilde{h}) + c_{q,q'}, z^{\text{sp}}(q, \tilde{h}+1)\}$. For each state we now have all the constrained lower bound values available that correspond to $b_i^{\text{CVRPC}}(s, \tilde{h})$. Additionally, we define $b_i^{\text{CVRPC}, \geq}(s, \tilde{h}) = \min_{\tilde{h}' \in \{\tilde{h}, \ldots, \tilde{h}_i^{\max}\}} b_i^{\text{CVRPC}}(s, \tilde{h}') \, \forall \tilde{h} \in \{\tilde{h}_i^{\min}, \ldots, \tilde{h}_i^{\max}\}$, which gives us the lower bounds when using at least $\tilde{h}$ home stands.

Table 1. Memory demand for different lower bound lookup tables over the number of teams in GB assuming two bytes per bound value.

n	TSP	CVRP	CVRPH
14	0.003	0.009	0.127
16	0.016	0.047	0.75
18	0.079	0.237	4.27
20	0.390	1.172	23.43

Table 2. Runtimes in minutes for CVRPH bound calculations for NL14 to NL16 and CIRC14 to CIRC18.

	14	16	18
NLn	25	169	–
CIRCn	25	173	903

In Table 1, we see the memory demand for the three different lower bound lookup tables in GB assuming 2 bytes per bound value. Up to 16 teams, they all have reasonable size and our experiments have shown that the 16 teams instance bounds can be pre-calculated within three hours in a prototypic Python 3.7 implementation on an Intel Xeon E5-2640 processor with 2.40 GHz in single-threaded mode, see Table 2. 18 teams instances are also within reach with the strong CVRPH bound taking already 15 h—we suppose that an order of magnitude in time can be saved using a compiled language. For larger instances, these numbers and the computation times increase dramatically, since the number of bounds grows for the CVRPH bound with $\mathcal{O}(n^3 2^n)$—already the TSP bound, being the weakest, needs 42 GB for 26 teams.

For even a further tightening of the CVRPH bound, we make use of the minimum number of trips (MNT) bound by Urrutia et al. [10]. It does not assume strong independence between the teams anymore. Instead, the relaxed CONSTANT variant of the problem, where all distances are set to one is solved to optimality (or taking a lower bound), yielding a minimum number of trips all teams together have to perform in a feasible solution of the problem. A trip in this case is an atomic movement of a team from one venue to another. Given a state s, let us call $\tau = \sum_{i=1}^{n} t_i$ the number of trips performed so far by all teams in the shortest path from s_r to s. By the CVRPC bound, each team has an optimal number $\tilde{h}_i^{\text{opt}}$ of home stands from s to s_t. This translates to an optimal number of trips $t_i^{\text{opt}} = |\mathcal{A}_i^s| + \tilde{h}^{\text{opt}} - 1_{x_i = i}$. Let τ^{lb} be the lower bound for the minimum number of trips. If $\tau^{\text{lb}} \leq \tau + \sum_i t_i^{\text{opt}}$, then we cannot tighten the CVRPH bound further. Otherwise, we can add the constraint that the teams have to perform $\Delta\tau = \tau^{\text{lb}} - \tau - \sum_i t_i^{\text{opt}}$ extra trips, yielding the MNT bound

$$b^{\mathrm{MNT}}(s) = \min \quad \sum_{i=1}^{n} b_i^{\mathrm{CVRPC},\geq}(s, \tilde{h}_i) \tag{8}$$

$$\text{s.t.} \quad \tilde{h}_i \in \{\tilde{h}_i^{\min}, \dots, \tilde{h}_i^{\max}\} \quad \forall i \in 1, \dots, n \tag{9}$$

$$\tau + \underbrace{\sum_{i=1}^{n} |\mathcal{A}_i^s| + \tilde{h}_i - 1_{x_i = i}}_{|P^s|} \geq \tau^{\mathrm{lb}} \tag{10}$$

This bound can be calculated by solving a corresponding integer linear program using binary decision variables $y_i^{\tilde{h}}$ with costs derived from the CVRPC lower bound function $c_i^{\tilde{h}}$ and counting values $d_i^{\tilde{h}} \in \{\tilde{h}_i^{\min}, \dots, \tilde{h}_i^{\max}\}$:

$$b^{\mathrm{MNT}}(s) = \min \quad \sum_{i=1}^{n} c_i^{\tilde{h}} y_i^{\tilde{h}} \tag{11}$$

$$\text{s.t.} \quad \sum_{\tilde{h}} y_i^{\tilde{h}} = 1 \quad \forall i \in 1, \dots, n \tag{12}$$

$$\sum_{i,\tilde{h}} y_i^{\tilde{h}} d_i^{\tilde{h}} \geq \tau^{\mathrm{lb}} - |P^s| - \tau + \sum_{i=1}^{n} 1_{x_i = i} \tag{13}$$

We take the τ^{lb} values from [7], where Rasmussen and Trick present a Benders decomposition approach to solve the CONSTANT instances for up to 16 teams each within at most five minutes.

7   Computational Study

We conducted all our experiments on Intel Xeon E5-2640 processors with 2.40 GHz in single-threaded mode and a memory limit of 32 GB. We implemented our approach as a prototype in Python 3.7, being aware that an implementation in a compiled language would likely be substantially faster and have a smaller memory footprint. To solve the integer linear programs for the MNT bound, we used Gurobi 12.8.

In Table 3, we present a comparison of the results obtained by the deterministic beam search with beam widths 1000 and 10000 for the instance sets NL and CIRC (See footnote 1) [4] over the different lower bounds used in the state ordering. We see that the weak TSP bound does not provide good guidance and even misguides the search for larger instances, where using no bound ($b(s) = 0$) and sorting the nodes only by the currently shortest path length to them (SHORT) provides better results. Much better guidance can be observed for the CVRP based bounds, where we see similar improvements over all instances.

Table 3. Final solution lengths of deterministic beam search with different state ordering heuristics and beam widths with lexicographic team orderings. Sorting the states by currently shortest path length to them (SHORT) does not use any lower bound; for a description of the TSP, CVRP, CVRPH, MNT lower bounds refer to Sect. 6. For circ16 with MNT we did not achieve a final result due to excessive runtime.

inst	$\beta = 1000$					$\beta = 10000$				
	SHORT	TSP	CVRP	CVRPH	MNT	SHORT	TSP	CVRP	CVRPH	MNT
nl6	24876	24759	23954	**23916**	**23916**	24876	**23978**	**23916**	**23916**	**23916**
nl8	42308	41977	40687	40687	40687	40970	41762	**39776**	**39776**	**39776**
nl10	67094	66469	62329	60713	62400	66087	64700	61129	60757	**60554**
nl12	131046	129209	116976	114499	114499	127238	119271	**113294**	114475	114824
nl14	217763	233765	211643	211116	211116	224537	219708	203519	**203279**	**203279**
nl16	309227	–	283985	285326	286085	301989	322567	276599	275562	**271251**
circ6	66	**64**	**64**	**64**	**64**	**64**	**64**	**64**	64	64
circ8	144	146	**134**	**134**	**134**	136	142	**134**	**134**	134
circ10	280	284	264	262	266	268	276	**246**	**246**	250
circ12	452	502	428	430	430	444	468	**418**	**418**	418
circ14	734	774	674	672	672	710	760	668	**656**	**656**
circ16	–	–	1012	1000	990	1012	1114	**956**	966	n/a

Since the number of classic benchmark instances is limited and to further validate the guidance quality of the different bounds, we created two different types of random instances. First, the set $\mathcal{I}_{L^1}$, where we sample 30 instances for team numbers 8, 10, and 12 each on an integer grid of size 1000×1000 using Manhattan distances to compute the resulting distance matrices; second, the set $\mathcal{I}_{L^2}$, using the same sampling procedure but using the rounded to the nearest even integer Euclidean distances. This yields in total 180 additional test instances. We exclude the TSP bound from our further experiments since it did not show promising results for the tests on the NL and CIRC instances. In Table 4, we see mean values of final solution lengths and corresponding standard deviations when performing the deterministic beam search with the different state ordering heuristics on the randomly generated instances. The gap between SHORT and CVRP is well observed especially with 10 and 12 teams. The gap between CVRP and CVRPH is closer, a Wilcoxon signed rank sum test shows that we can reject the assumption that CVRP is better than CVRPH with a significance level of $\alpha = 1\%$. The difference between CVRPH and MNT is for the L^1 distance instances inconclusive, and for the L^2 instances slightly in favor of MNT, but at the cost of substantially higher runtime due to the linear programs that need to be solved for every state. For further experiments we therefore limit ourselves to the CVRP/CVRPH bounds.

Finally, Table 5 compares our randomized beam search variant with either lexicographic or random team ordering performed in parallel and independently on 30 cores with several state-of-the-art approaches on three difficult NL and CIRC instances. Each beam search run was conducted with beam width $\beta = 10^5$ and randomization parameter $\sigma_{\mathrm{rel}} = 0.001$ resulting in equally gentle noise

Table 4. Comparison of our beam search algorithm with $\beta = 1000$ over different state ordering heuristics on 180 randomly generated test instances with 8, 10, and 12 teams, using Manhattan and Euclidean distances, evenly split. Mean values of final solution lengths and standard deviations over 30 test instances are shown.

class	$\beta = 1000$			
	SHORT	CVRP	CVRPH	MNT
$\mathcal{I}_{L^1}^{8}$	42532 ± 5384	40530 ± 5214	40405 ± 5030	40405 ± 5030
$\mathcal{I}_{L^1}^{10}$	70049 ± 7280	65483 ± 6886	64760 ± 6689	64964 ± 6922
$\mathcal{I}_{L^1}^{12}$	99086 ± 7991	92838 ± 8089	91728 ± 7726	91465 ± 7694
$\mathcal{I}_{L^2}^{8}$	34412 ± 5088	33034 ± 5109	32965 ± 5071	32965 ± 5071
$\mathcal{I}_{L^2}^{10}$	55019 ± 5872	51723 ± 5988	51269 ± 5808	51057 ± 5829
$\mathcal{I}_{L^2}^{12}$	79699 ± 7293	74231 ± 6933	73700 ± 6456	73524 ± 6403

applied to the f-values of the states in every layer. The noise parameter was determined using irace [5] on the randomly generated instances. The table shows minimum and mean values for solution lengths of finally best solutions. We observe that we can compete well with the other mainly constructive approach "ant colony optimization with forward checking and conflict-directed backjumping" (AFC-TTP) from [11] and the composite-neighborhood tabu search (CNTS) from [3] on the NL instances and obtain better results than these for the CIRC instances, without hybridizing with a final local search. For CIRC instances we can also obtain similar results to population-based simulated annealing from scratch (PBSAFS) from [14], which uses parallel simulated annealing. For the circular instances with 14 and 16 teams, we found new best feasible solutions, as

Table 5. Comparison of the final solution lengths of parallel randomized beam search using either lexicographic team ordering or random team ordering (RTO) with 30 independent runs each, parameters $\sigma_{rel} = 0.001$, $\beta = 10^5$, and the CVRPH lower bound function (RBS-CVRPH) with the reported solution lengths of ant-colony optimization (AFC-TTP) [11], composite-neighborhood tabu search (CNTS) [3], simulated annealing (TTSA) [1], and population-based simulated annealing (PBSA) [14], where the latter is either used from scratch (PBSAFS) or starting from an already high quality solution (PBSAHQ) provided by a TTSA run. [†]New best feasible solutions.

inst	RBS-CVRPH		RBS-CVRPH-RTO		AFC-TTP		CNTS		TTSA		PBSAFS		PBSAHQ	
	min	mean	min	mean	min	mean	min	mean	min	mean	min	mean	min	mean
nl12	112680	113594.6	112791	113581.5	112521	114427.4	113729	114880.6	112800	113853.0	**110729**	**112064.0**	n/a	n/a
nl14	192625	198912.6	196507	199894.8	195627	197656.6	194807	197284.2	190368	192931.9	**188728**	190704.6	**188728**	**188728.0**
nl16	266736	271367.1	265800	270925.9	280211	283637.4	275296	279465.8	267194	275015.9	**261687**	265482.1	262343	**264516.4**
circ12	410	415.7	410	414.6	430	436.0	438	440.4	n/a	n/a	**404**	418.2	408	414.8
circ14	632	641.0	630[†]	640.7	674	692.8	686	694.4	n/a	n/a	640	654.8	632	645.2
circ16	918	933.8	910[†]	931.6	1034	1039.6	1016	1030.0	n/a	n/a	958	971.8	916	917.8
circ18	1300	1322.0	1296	1320.4	1486	1494.8	1426	1440.8	n/a	n/a	1350	1371.6	**1294**	1307.0

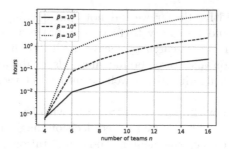

Fig. 3. Runtimes in hours for deterministic beam search runs on NL instances with $\beta \in \{10^3, 10^4, 10^5\}$.

of the time of writing according to Michael Trick's TTP web page. The strongest results overall for NL and CIRC are provided by simulated annealing (TTSA) from [1] and its parallel variant PBSA from [14].

Runtimes of our approach are shown in Fig. 3 measured for deterministic beam search on the NL instances up to 16 teams for $\beta \in \{10^3, 10^4, 10^5\}$. For example, a run on an instance with 12 teams and a beam width of 10^5 takes roughly 10 h. We believe it is possible to improve this further by an order of magnitude using a compiled language.

8 Conclusion and Future Work

We investigated a beam search approach for the well-known traveling tournament problem. To this end, we proposed a recursive state space formulation, which is searched by a restricted breadth-first-search. This beam search is implemented in a memory efficient variant allowing for high beam widths to be tested. For guiding the search, we studied different lower bounds derivable from a state. Furthermore, we introduced a randomized beam search variant which applies parameterized Gaussian noise to the state ordering heuristic in order to diversify the search when performing multiple runs in parallel. We contribute a method based on decision diagrams to pre-calculate the existing capacitated vehicle routing problem and minimum number of trips bounds for instances up to 18 teams and how these bounds can be effectively used for any given state. To compare different lower bounds and tune algorithmic parameters, we created artificial instances. This allowed us to ultimately achieve better results on difficult NL and CIRC benchmark instances than the also mainly constructive ant-colony optimization approach AFC-TTP and the composite-neighborhood tabu search CNTS. For the circular instances with 14 and 16 teams we could find new best feasible solutions. Overall, the simulated annealing based approaches TTSA/PBSA still remain dominant.

We have implemented our approach as a prototype in Python 3.7. A re-implementation in a compiled language is desirable as much better runtimes and smaller memory footprints can be expected, which would allow to tackle even higher beam widths. So far we did not consider any local search, but a natural extension would be to try to further improve a number of best solutions provided by our beam search by local search. Furthermore, the provided state space formulation and lower bound methods might also be incorporated into GRASP or ACO algorithms, as well as into exact techniques such as A* variants.

To tackle instances with more than 18 teams with lower bound guidance, an interesting direction could be to use relaxed decision diagram for the bound pre-calculations, in order to keep the memory and computational demand reasonably bounded.

References

1. Anagnostopoulos, A., Michel, L., Van Hentenryck, P., Vergados, Y.: A simulated annealing approach to the traveling tournament problem. J. Sched. **9**(2), 177–193 (2006)
2. Bergman, D., Cire, A.A., van Hoeve, W.J., Hooker, J.N.: Decision Diagrams for Optimization. AIFTA. Springer, Cham (2016). https://doi.org/10.1007/978-3-319-42849-9
3. Di Gaspero, L., Schaerf, A.: A composite-neighborhood tabu search approach to the traveling tournament problem. J. Heuristics **13**(2), 189–207 (2007)
4. Easton, K., Nemhauser, G., Trick, M.: The traveling tournament problem description and benchmarks. In: Walsh, T. (ed.) CP 2001. LNCS, vol. 2239, pp. 580–584. Springer, Heidelberg (2001). https://doi.org/10.1007/3-540-45578-7_43
5. López-Ibáñez, M., Dubois-Lacoste, J., Cáceres, L.P., Birattari, M., Stützle, T.: The irace package: iterated racing for automatic algorithm configuration. Oper. Res. Perspect. **3**, 43–58 (2016)
6. Ow, P.S., Morton, T.E.: Filtered beam search in scheduling. Int. J. Prod. Res. **26**(1), 35–62 (1988)
7. Rasmussen, R.V., Trick, M.A.: A benders approach for the constrained minimum break problem. Eur. J. Oper. Res. **177**(1), 198–213 (2007)
8. Ribeiro, C.C., Urrutia, S.: Heuristics for the mirrored traveling tournament problem. Eur. J. Oper. Res. **179**(3), 775–787 (2007)
9. Thielen, C., Westphal, S.: Complexity of the traveling tournament problem. Theoret. Comput. Sci. **412**(4–5), 345–351 (2011)
10. Urrutia, S., Ribeiro, C.C., Melo, R.A.: A new lower bound to the traveling tournament problem. In: 2007 IEEE Symposium on Computational Intelligence in Scheduling, pp. 15–18. IEEE (2007)
11. Uthus, D.C., Riddle, P.J., Guesgen, H.W.: An ant colony optimization approach to the traveling tournament problem. In: Proceedings of the 11th Annual Conference on Genetic and Evolutionary Computation, pp. 81–88. ACM (2009)
12. Uthus, D.C., Riddle, P.J., Guesgen, H.W.: DFS* and the traveling tournament problem. In: van Hoeve, W.-J., Hooker, J.N. (eds.) CPAIOR 2009. LNCS, vol. 5547, pp. 279–293. Springer, Heidelberg (2009). https://doi.org/10.1007/978-3-642-01929-6_21
13. Uthus, D.C., Riddle, P.J., Guesgen, H.W.: Solving the traveling tournament problem with iterative-deepening A*. J. Sched. **15**(5), 601–614 (2012)
14. Van Hentenryck, P., Vergados, Y.: Population-based simulated annealing for traveling tournaments. In: Proceedings of the 22nd National Conference on Artificial Intelligence, no. 1, pp. 267–262. MIT Press (2007)

Cooperative Parallel SAT Local Search
with Path Relinking

Padraigh Jarvis[1] and Alejandro Arbelaez[2(✉)]

[1] United Technologies Research Centre, Cork, Ireland
JarvisPa@utrc.utc.com
[2] School of Computer Science and Information Technology,
Insight Centre for Data Analytics, University College Cork, Cork, Ireland
a.arbelaez@cs.ucc.ie

Abstract. In this paper, we propose the use of path relinking to improve the performance of parallel portfolio-based local search solvers for the Boolean Satisfiability problem. In the portfolio-based framework several algorithms explore the search space in parallel, either independently or cooperatively with some communication between the solvers. Path relinking is a method to maintain an appropriate balance between diversification and intensification (and explore paths that aggregate elite solutions) to properly craft a new assignment for the variables to restart from. We present an empirical study that suggest that path relinking outperforms a set of well-known parallel portfolio-based local search algorithms with and without cooperation.

Keywords: SAT · Parallel local search

1 Introduction

The propositional satisfiability problem (SAT) is a fundamental problem in computer science with important applications ranging from bioinformatics [19] to planning [23] and scheduling [22]. The SAT problem consists in determining whether a given Boolean formula $\mathcal{F}$ is satisfiable or not. This formula is usually represented using the *Conjunctive Normal Form* (CNF) as follows: $\mathcal{F} = \bigwedge_i \bigvee_j l_{ij}$, where each l_{ij} represents a literal (a propositional variable or its negation) and the disjunctions $\bigvee_j l_{ij}$ are the clauses in $\mathcal{F}$. A k-SAT problem indicates that $\mathcal{F}$ contains k literals per clause, for instance a 3-SAT formula can be represented as follows:

$$\mathcal{F} = (v_{11} \vee v_{12} \vee v_{13}) \wedge (v_{21} \vee v_{22} \vee v_{23}) \dots (v_{n1} \vee v_{n2} \vee v_{n3})$$

In the weighted MaxSAT problem, clauses are associated with a positive weight and the problem consists in minimizing the cost, i.e., the sum of weights of unsatisfied clauses. The weighted partial MaxSAT problem consists in finding a solution (or an assignment for the variables) that minimizes cost while satisfying a given subset of clauses (i.e., hard clauses).

© Springer Nature Switzerland AG 2020
L. Paquete and C. Zarges (Eds.): EvoCOP 2020, LNCS 12102, pp. 83–98, 2020.
https://doi.org/10.1007/978-3-030-43680-3_6

Complete parallel solvers for the SAT problem have received significant attention recently, these solvers can be divided into two categories the classical divide-and-conquer approach [14] and the parallel portfolio approach [1,7]. The first one typically divides the search space into several sub-spaces, and the second one lets algorithms compete and cooperate to solve a given problem instance.

In this paper, we focus our attention in cooperative parallel local search solvers for the SAT and Weighted Partial MaxSAT problems. In our settings, each member of the portfolio shares its best assignment for the variables. At each restart point, instead of classically generating a random assignment to start with, the portfolio aggregates the shared knowledge to carefully craft a new starting point.

This paper is organized as follows. Section 2 presents key concepts of local search, including a description of a set of well-known variable selection methods to tackle SAT and MaxSAT problems. Section 3 provides general concepts about parallel portfolios of local search algorithms. Section 4 describes our new cooperative policies using path relinking. Section 5 evaluates our new cooperative policies and Sect. 6 presents concluding remarks and areas of future work.

2   Local Search for SAT and MaxSAT

Algorithm 1 describes the general schema of the local search procedure for the SAT problem. It starts with a random assignment for the variables in the formula F (*initial-solution* line 2). The key point of the local search procedure is depicted in lines 3–9 where the algorithm flips the most appropriate variable until a certain stopping condition is met, e.g., a given number of flips is reached (Max-Flips) or after a given timeout. After this procedure the algorithm restart itself with a new fresh random assignment for the variables.

Algorithm 1. Local Search (CNF formula F, Max-Flips, Max-Tries)

1: **for** try := 1 to Max-Tries **do**
2:     A := initial-solution(F)
3:     **for** flip := 1 to Max-Flips **do**
4:         **if** A satisfies F **then**
5:             return A
6:         **end if**
7:         x := select-variable(A)
8:         A := A with x flipped
9:     **end for**
10: **end for**
11: return 'No solution found'

As one may expect, a critical part of the algorithm is the variable selection function (line 7 *select-variable*), which indicates the next variable to be flipped in the current iteration of the algorithm. Currently, nearly all variable selection algorithms are variations of the GSAT [18] and WalkSAT [17] algorithms originally proposed for the SAT problem. These two algorithms attempt to select the variable with the highest *score*.

$$score(x) = make(x) - break(x)$$

Intuitively, *make(x)* indicates the total number of clauses that are currently unsatisfied but become satisfied after flipping x. Similarly, *break(x)* indicates the total number of clauses that are currently satisfied but become unsatisfied after flipping x. Taking this into account, local search algorithms tend to select variables with the minimum *score*, flipping those variables would most likely increase the chances of obtaining the optimal assignment for the variables. In the following, we describe seven well-known variable selection algorithms for the SAT and MaxSAT problems.

- *WalkSAT* [17] uniformly at random selects an unsatisfied clause c. Then, with a probability *wp* selects a random variable from c and with probability *1-wp* identifies the most suitable variable in c.
- *AdaptNovelty+ (AN+)* [8] uses an adaptive mechanism to properly self-tune the noise parameter (wp) of WalkSAT algorithms (e.g., *Novelty+*).
  AdaptNovelty+ introduces a new parameter ϕ to control the value of *wp*. *wp* is initially set to 0 and updated when search stagnation is observed, i.e., $wp = wp + (1+wp) \times \phi$. Additionally, whenever an improvement is observed *wp* is decreased, i.e., $wp = wp - wp \times \phi/2$. The authors define search stagnation as a stage when no improvement has been observed in the objective function for a given number of iterations.
- *G2WSAT* (G2) [11] introduces the concept of promising decreasing variable. Broadly speaking, a variable is decreasing if flipping it improves the objective function (i.e., total number of (weighted) violated clauses).
- *Adaptive G2WSAT* (AG2) [11] aims to integrate an adaptive noise mechanism into the G2WSAT algorithm.
- *PAWS* (Pure Additive Weighting Scheme) [20] assigns a weight penalty to each clause, with those that go unsolved having their weight penalty changed. This solver includes a chance to make a flip that will result in a lateral movement in satisfiability and a variable to determine how often the weights of clauses are changed.
- *Dist* [5] proposes a variable selection scheme based on hard and soft clauses. *Dist* initially, maintains a list of hard-decreasing variables (i.e., a set of promising decreasing variables of hard clauses), and the algorithm defines a hard and a soft score for the variables in the problem. Furthermore, the authors propose to on-the-fly adjust the weight of hard clauses. This way, *Dist* bias the variable selection process towards improving the score of hard clauses with the set of hard-decreasing variables.
- *CCLS* [13] maintains a list of candidate variables CCMPVars (Configuration Checking and Make Positive), each variable x in the list has a make(x) > 0 and the age of x is smaller than at least one of its neighbour variables (i.e., a variable sharing at least one clause). In the diversification phase with

probability p performs a random walk step; otherwise, with a probability 1-p in the intensification phase the algorithm selects the variable with the greatest score in CCMPVars. However, if CCMPVars is empty the algorithm performs a random walk. *CCEHC* [12] extends *CCLS* to prioritize the search towards variables involved in violated hard clauses.

3   Parallel Local Search

In this paper, we use the traditional parallel portfolio framework by executing several algorithms in parallel (or different copies of the same one with different random seeds). Therefore, each algorithm independently executes a sequential restart-based local search algorithm and we periodically restart the algorithms to aggregate the common knowledge of the portfolio.

$$
M = \begin{pmatrix} X_{11} & X_{12} & \ldots & X_{1n} \\ X_{21} & X_{22} & \ldots & X_{2n} \\ \vdots & \vdots & \vdots & \vdots \\ X_{c1} & X_{c2} & \ldots & X_{cn} \end{pmatrix}
$$

Fig. 1. Pool of elite solutions.

In our parallel algorithm we maintain a pool of elite solutions. In this context, each algorithm in the portfolio shares the best solution observed so far in a shared pool M (see Fig. 1). Where n indicates the total number of variables of the problem and c indicates the number of local search algorithms in the portfolio. In the following we are associating local search algorithms and processing cores. Each element X_{ji} in the pool denotes the ith variable of the best solution found so far by the jth core.

The initial restarting solution of the algorithms in the portfolio is determined by the cooperation protocol and is a composition of the solutions in the pool. Therefore, maintaining an appropriate balance between diversification and intensification of the solutions in the pool is an important step in the proposed cooperative framework. We remark that we use a random solution for the first start and the cooperative framework afterwards.

Recently, [2] proposed seven cooperative algorithms for parallel SAT solving. These strategies range from a voting mechanism, where each algorithm in the portfolio suggest a value for each variable, to probabilistic constructions. This way, the *variable-initialization* function (line 2, Algorithm 1) uses cooperation (after the second restart) in lieu of random values for the variables.

Prob uses a probability function based in the number of occurrences of variables with positive and negative values. *PNorm* normalizes the probability function with the quality of the solutions (i.e., number of unsatisfied clauses), therefore, values involved in better truth assignments are most likely to be used in the future. Complete details about these two popular cooperative techniques are available in [2].

Other work in the area includes PGSAT [15], a parallel version of the GSAT algorithm. The entire set of variables is randomly divided into τ subsets and allocated to different processors. In this way at each iteration, if no global solution has been obtained, the ith processor uses the GSAT score function to select and flip the best variable for the ith subset. Another contribution to this parallel architecture is described in [16] where the authors aim to combine PGSAT and random walk. Thus at each iteration, the algorithm performs a random walk step with a certain probability wp, that is, a random variable from an unsatisfied clause is flipped. Otherwise, PGSAT is used to flip τ variables in parallel at a cost of reconciling partial configurations to test if a solution is found.

4   Path Relinking

Path relinking [6] is a popular technique to generate new solutions by exploring paths that connect elite solutions. To generate the new solution (i.e., line 2 in Algorithm 1), an initial solution and a guiding solution are selected from the pool to represent the starting and the ending points of the path.

Path of solutions

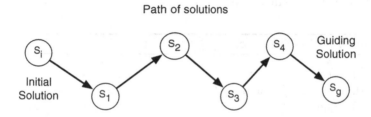

Fig. 2. Path relinking.

In this paper, we use path relinking to generate new starting solutions for the algorithms in the portfolio. Figure 2 depicts the process of generating the path of solutions. We select the initial (s_i) and guiding (s_g) solutions, and then the path relinking algorithm generates the intermediate solutions by replacing values for the variables in s_i with values from s_g. In the context of this paper, the path of solutions regulates the intensification/diversification trade-off. The first neighbour solution of s_i denotes, at least, one change in the initial solution and n-1 chances in the guiding solution.[1] Ideally, in order to balance the diversification/intensification trade-off, the new generated solution should be in middle between the s_i and s_g.

Algorithm 2 shows our path relinking algorithm to generate a new starting solution. Let s_i and s_g denote the initial and guiding solutions and $pd \in [0, 1]$

---

[1] n denotes the number of variables in the problem.

denotes the probability of using the initial or guiding solution for each variable in the problem. In particular, we explore the following four alternatives to define s_i and s_g:

- *best2rand (b2r)*: s_i represents the best solution in M and s_g is randomly selected from M;
- *cbest2rand (cb2r)*: s_i represents the best solution obtained so far for the processing unit that is currently seeking a new starting solution and s_g is randomly selected from M;
- *best2cbest (b2cb)*: s_i represents the best solution in M and s_g represents the best solution obtained so far for the processing unit that is currently seeking a new starting solution;
- *best2sbest (b2sb)*: s_i represents the best solution in M and s_g represents the second best solution available in M.

The path relinking algorithm uses pd to balance the diversification vs. intensification trade-off dilemma, a value close to 1 (resp. 0) favours s_g (resp. s_i). Therefore, $pd = 0.5$ is a reasonable value for a proper intensification/diversification balance of the solutions. Furthermore, *best2rand* and *cbest2rand* provide further diversification benefits as the method randomizes the selection of the solutions in the pool. Certainly, biasing the search towards s_i or s_g might improve performance for specific problem families. However, without explicit knowledge of the target instances we recommend $pd = 0.5$.

Algorithm 2. Path-relinking(S_i, S_g, pd)

1: $s := s_i$
2: **for** i := 1 to $|s|$ **do**
3:    **if** with probability pd **then**
4:        $s[i] := s_g[i]$
5:    **end if**
6: **end for**
7: **return** s

---

5   Experiments

In this section, we present experiments for our cooperative parallel portfolios using path relinking for SAT and Weighted Partial MaxSAT Solving. We decided to build our parallel portfolio on top of UBCSAT [21], a well-known local search library that provides efficient implementations of popular local search algorithms for SAT and MaxSAT.

In our experiments we use the sequential local search algorithms with their default parameters and MaxFlips = 10^6 except for non-cooperative algorithms. Indeed, sequential algorithms are equipped with important diversification techniques and usually perform better without restarts and therefore we use MaxFlips = ∞ for non-cooperative parallel portfolios.

5.1 SAT Experiments

In these experiments we consider all known satisfiable uniform random k-SAT instances from the 2017 and 2018 SAT competitions (for a total collection of 174 instances).[2] and we consider the following algorithms: *AN+*, *G2WSAT*, *PAWS*, *AG2*. We evaluated the impact of our cooperative policies with two versions of the portfolio. The first version analyses the impact of the new policies with multiple copies of the same algorithm and the second one implements a parallel portfolio with four different algorithms.

We conducted this set of experiments on 15 machines running Ubuntu18.04 with 16 GB of RAM and a AMD Razen 5 2400 g CPU with 4 cores. We ran each solver 5 times on each instance (each time with a different random seed) with a 5-min time cutoff. For each pair ⟨instance, solver⟩ we compute the median time and the Penalized Average Runtime (*PAR10*), i.e., average runtime, but unsolved instances are considered as 10× the time-limit [9], over all 5 runs.

Figure 3 shows the cactus plot of the parallel portfolios with multiple copies of the same algorithm. The y-axis gives the number of solved instances and the x-axis presents the cumulative runtime. In this figure, it can be seen that our new cooperative policies with path relinking outperform (except for *AN+*) existing techniques such as: *PNorm*, *Prob*, and a portfolio without cooperation (*non-coop*).

Figure 3(a) and (b) show the performance of the two weakest algorithms, that is, *AG2* with 51 instances in 280 s (for *Prob*) and *AN+* with 43 instances in 267 s (for *best2cbest*). Figure 3(c) shows that *best2cbest* is the best cooperative policy for *PAWS* solving 60 instances in 276 s. On the other hand, Fig. 3(d) summarizes the performance of the best algorithm, it can be observed that the non-cooperative framework outperforms the other methods when the time cutoff is up 150 s. However, after this point, *best2sbest* and *best2cbest* report outstanding performances with respectively 74 instances in 275 s and 72 instances in 281 s.

Figure 4 shows the cactus plot of the parallel portfolio with different algorithms, *sequential* reports the performance of the best sequential algorithm (i.e., *G2WSAT*). Similarly to Fig. 3(d) the non-cooperative portfolio reports a very good performance up to about 200 s. However, after this point our new path relinking policies largely outperform *non-coop*, *Prob* and *PNorm*. In particular, *cbest2rand* (resp. *best2cbest*) solves 7.4% (resp. 12%) more instances than *PNorm* (resp. *Prob*).

Table 1 reports complete details of the performance of a 4-core portfolio with different algorithms. # Solved reports the number of solved instances within the time limit, Time denotes the average time in seconds for solved instances (i.e., average across instances of the median across 5 runs on a given instance), and PAR10 reports the average PAR10 of the parallel portfolio. It can be observed that all our new path relinking policies outperform existing methods (except *best2sbest*), i.e., *cbest2rand* and *best2cbest* solve seven (resp. five) more instances than *PNorm* (resp. Non-cooperation).

---

[2] https://satcompetition.org/.

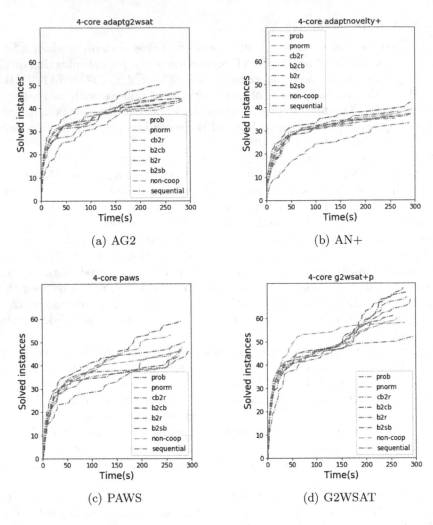

Fig. 3. Cactus plot for 4-core portfolios using copies of the same algorithm.

Table 1. Portfolio full results

Method	# Solved	Time	PAR10
Sequential	53	51	2101
Non-cooperative	63	41	1928
Pnorm	65	58	1901
Prob	62	55	1950
best2cbest	70	73	1822
cbest2rand	**70**	**69**	**1821**
best2rand	68	66	1853
best2sbest	64	61	1919

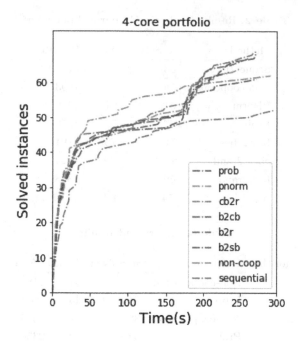

Fig. 4. Cactus for 4-core portfolio with different algorithms

Table 2 reports the performance of a 4-core parallel portfolio with multiple copies of the best sequential algorithm, i.e., *G2WSAT*. In this experiment, we observe that all our new policies outperform well-known techniques, for instance, the worst path relinking algorithm solves more instances than the best exiting technique for this experiment (i.e., *Prob*). Furthermore, our overall best new policy (i.e., *best2sbest*) solves 21% more instances than the non-cooperative portfolio and 25% more instances than the best existing method. As expected the time of solved instances increases as this parallel algorithm solves more instances.

Finally, Table 3 summaries the overall results of the 4-core portfolios. Algorithm indicates the base local search algorithm. Coop. Policy indicates the cooperative policy, each cell shows the performance of Non-cooperative portfolios, Prob, and the best parallel portfolio for the reference algorithm.[3] As it can be observed the cooperative portfolio always outperform the non-cooperative one, and our new path relinking policies outperform all other policies in 4 out of 5 experiments. Furthermore, *G2* equipped with best2sbest is the overall winner policy with 74 solved instances in 87 s.

---

[3] Please notice that *AG2* only reports two cooperative policies as Prob is the winner strategy.

Table 2. Results for 4-core G2 parallel portfolios

Method	# Solved	Time	PAR10
Sequential	53	51	2101
Non-cooperative	59	31	1993
PNorm	61	57	1968
Prob	62	55	1950
best2cbest	72	79	1791
cbest2rand	70	78	1824
best2rand	69	82	1843
best2sbest	**74**	87	**1761**

Table 3. Experiment results

Algorithm	Coop. policy	# Solved	Time	PAR10
AG2	Non-cooperation	48	54	2187
	Prob	51	45	2134
AN+	Non-cooperation	38	51	2355
	Prob	38	52	2356
	best2cbest	43	54	2272
G2	Non-Cooperation	59	31	1993
	Prob	62	55	1950
	best2sbest	**74**	**87**	**1761**
PAWS	Non-Cooperation	54	67	2090
	Prob	51	63	2139
	best2cbest	60	41	1990
Portfolio	Non-Cooperation	63	41	1928
	Prob	62	55	1950
	cbest2rand	70	69	1821

5.2   Weighted Partial MaxSAT

We conducted experiments using crafted and random instances. The first dataset is a collection of 234 crafted instances used regularly in the annual MaxSAT competitions: staff-scheduling (12), auctions/auc-paths (20), auctions/auc-scheduling (20), min-enc/planning (30), warehouses (18), casual-discovery (35), csg (10), random-net (32), set-covering (45), mip-lib (12).

For the second dataset, we followed a similar approach as [13] and used *makewff* [24] to generate 270 uniform random weighted partial MaxSAT instances around the phase transition, i.e., 90 3-SAT instances, 90 5-SAT instances, and 90 7-SAT instances; and the number of variables per instance ranges from 2000 to 4000 (3-SAT), 1000 to 3000 (5-SAT), and 300 to 500

(7-SAT). For each random instance we randomly split clauses into two disjoint sets with hard and soft clauses. The number of hard clauses varies between 10%–40% of the total clauses in the problem.

After a preliminary experimentation we decided to use multiple copies of *AdapNovelyt+* to build our parallel portfolio for the random dataset. We would like to remark that *WalkSAT* and *G2WSAT* reported a poor performance and were unable to find feasible solutions for this problem family (i.e., satisfying all hard clauses). Alternatively, we use *AdaptNovelty+*, *WalkSAT*, and *G2WSAT* for crafted instances, so that we build our portfolios for crafted instances as follows:

- 4 Cores: *AdaptNovelty+* (2 cores), *WalkSAT* (1 core), and *G2WSAT* (1 core);
- 8 Cores: *AdaptNovelty+* (3 cores), *WalkSAT* (2 cores), and *G2WSAT* (3 cores).

We compare our cooperative algorithm against the following state-of-the-art local search solvers (with their recommended parameters): *Dist*, *CCEHC*, *CCLS*, *Prob*, and *PNorm*.[4] We remark that we use the same configurations for all our portfolios using the UBCSAT library. Unfortunately, *Dist*, *CCEHC*, *CCLS* do not support parallelism, and therefore, the only feasible parallel option for these solvers is the parallel portfolio without cooperation.

We conducted this set of experiments in the Microsoft Azure Cloud using DS4_v2 virtual machines with 28 GB of RAM and 8 cores at 2.40 Ghz Intel Xeon Processors E5-2673 running ubuntu. We ran each solver 5 times on each instance (each time with a different random seed) with a 5-min wall-clock timeout (300 s) for each experiment. For each pair ⟨instance, solver⟩ we compute the median time and solution quality over all 5 runs. Furthermore, we report the number of instances a given solver finds the best solution among all the solvers. We restart our local search solveres in all cooperative portfolios (i.e., *PNorm*, *best2rand*, and *best2cbest*) every 10^6 iterations or flips.

We start our evaluation with Fig. 5, we compare the performance of *best2rand* (cooperative portfolio) vs. *Portfolio* (non-cooperative portfolio) with eight cores. In both cases we use the same reference algorithms to build the portfolio.

As it can be seen in the figure, the cooperative framework implementing path relinking helps to considerably improve performance. For random instances (Fig. 5(a)) *best2rand* outperforms *Portfolio* for 128 instances; for 22 instances both solvers report the same solution cost; and only for 28 instances *Portfolio* outperforms *best2rand*. Alternatively, for crafted instances (Fig. 5(b)) *best2rand* outperforms *Portfolio* for 73 instances; *Portfolio* outperforms *best2rand* for 106 instances; and interestingly the non-cooperative portfolio only outperforms the cooperative one for 10 instances. It is also worth mentioning that *best2rand* is faster than *Portfolio* when both parallel solvers report the same solution cost, i.e., 13 and 47 times faster for random and crafted instances.

---

[4] In this paper, we use the implementation of *Dist*, *CCEHC*, *CCLS*, *Prob* and *PNorm* available in the SAT competitions and the website of the authors.

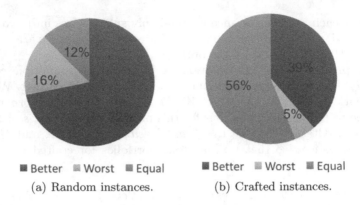

■ Better ■ Worst ■ Equal   ■ Better ■ Worst ■ Equal

(a) Random instances.   (b) Crafted instances.

Fig. 5. *best2rand* vs. *Portfolio*. Proportion of instances where the *best2rand* is better (resp. worst and equal) than *Portfolio* (counterpart portfolio without cooperation).

Table 4 presents further experimental results with the performance of sequential and parallel algorithms with and without cooperation using 4 and 8 cores for random instances. We recall that for random instances we build our parallel portfolio with and without cooperation using *AdaptNovelty+* (denoted as *Portfolio* in the table), and we omit the performance of *WalkSAT* and *G2WSAT* as these two solvers are unable to find feasible solutions for this problem family. Actually, the sequential version of the solvers in UBCSAT are unable to find the best solution for random instances (i.e., an assignment for the variables that satisfies all hard clauses), while *CCLS* and *CCEHC* solve 2 instances.

These results confirm that the cooperative approach with path relinking outperforms its counterpart portfolio with existing cooperative policies and without cooperation. For instance, *best2rand* solves (with 8 cores) respectively 36 and 32 more instances than the reference portfolio without cooperation (i.e., *Portfolio*) and *PNorm*.

As expected our non-cooperative portfolio is considerably weaker than *CCLS* (best sequential solver). However, adding our suggested cooperative framework leads to substantial performance improvements. As a result of that, our cooperative portfolio greatly outperforms the parallel version of *CCLS*, e.g., *best2rand* solves 7 and 16 more instances than *CCLS* with 4 and 8 cores.

We now switch our attention to crafted instances (Table 5). In this experiment, we include experimental results for our reference sequential solvers from UBCSAT (i.e., *WalkSAT*, *G2WSAT*, and *AdaptiveNovelt+*) as these solvers report competitive performance against modern local search solvers (i.e., *CCLS*, *Dist*, and *CCEHC*) for the Weighted Partial MaxSAT problem. In this dataset, it can be observed that *CCEHC* is the best sequential solver, reporting 109 instances with the best performance, followed by *Dist* (88 instances), and *AdaptiveNovelty+* (74 instances).

Similarly to random instances, our cooperative solver with path relinking outperforms its counterpart solvers *PNorm* (cooperative solver) and *Portfolio*

Table 4. Results for random instances.

Algorithm	Sequential		4 cores		8 cores	
	Time (s)	Best	Time (s)	Best	Time (s)	Best
best2rand	–	0	**189.6**	**22**	**157.4**	**54**
best2cbest	–	0	176.2	20	185.5	49
PNorm	–	0	149.8	7	162.9	22
Portfolio	–	0	114.1	6	166.3	18
CCLS	**183.9**	**2**	143.3	15	139.6	38
Dist	2.3	1	108.1	7	108.4	14
CCEHC	266.5	2	116.1	7	125.4	16

Table 5. Results for crafted instances.

Algorithm	Sequential		4 cores		8 cores	
	Time (s)	Best	Time (s)	Best	Time (s)	Best
best2rand	49.2	74	56.6	113	57.3	123
best2cbest	49.2	74	51.3	110	**87.0**	**138**
PNorm	49.2	74	38.8	101	38.4	103
Portfolio	49.2	74	35.4	98	26.2	99
AdaptNovelty+	49.2	74	40.8	86	42.9	91
WalkSAT	10.2	41	12.6	44	20.3	48
G2WSAT	25.8	57	16.1	60	12.1	61
CCLS	4.5	45	11.3	50	9.7	50
Dist	14.7	88	11.4	94	27.3	108
CCEHC	**15.1**	**109**	**13.0**	**115**	18.7	121

(parallel portfolio without cooperation). For instance, *bets2rand* solves 9 and 12 more instances than *PNorm* and *Portfolio* with 8 cores. *CCEHC* is the best portfolio with 4 cores solving 115 instances, 2 more than *best2rand*. This performance difference is mainly because *CCEHC* is considerably better (for this dataset) than our sequential algorithms from the UBCSAT library. Finally, *best2cbest* leads the ranking (with 8 cores) by solving 17 more instances than the parallel portfolio with the best sequential solver (i.e., *CCEHC*). Certainly, this performance improvement comes from our path relinking cooperative framework.

Finally, Figs. 6 and 7 show the cactus (i.e., number of solved instances with given a time limit) plot for 8-core portfolios for random and crafted instances. *best2rand* (b2r) is the best parallel solver, the second place is for *best2cbest* (b2c), and the third place is for *CCLS*. On the other hand, for crafted instances (Fig. 7) *best2cbest* and *best2rand* are the most effective solvers.

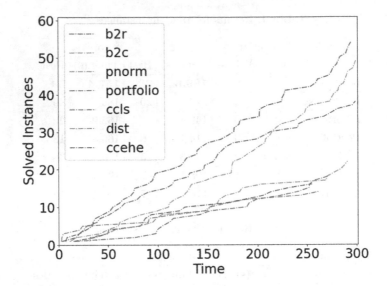

Fig. 6. Cactus plot for 8-core portfolios with random instances.

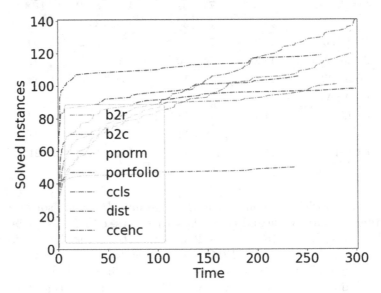

Fig. 7. Cactus plot for 8-core portfolios with crafted instances.

6   Conclusions and Future Work

In this paper, we proposed a cooperative framework using path relinking, a well-known technique to combine solutions in meta-heuristic search. The algorithm exploits parallelism by executing multiple local search algorithms in parallel, at each restart point, instead of classically generating a random solution to start with, we propose the use of path relinking to carefully craft new starting solutions.

Extensive experiments on a large number of instances for the SAT and Weighted Partial MaxSAT problems suggest that our new cooperative framework outperforms its counterpart parallel portfolio with and without cooperation. Furthermore, we have seen improvements for parallel portfolios with multiple copies of the same algorithm and parallel portfolios with different algorithms.

In the future, we would like to investigate the use offline and online tuning of the *pd* parameter to balance the diversification vs. intensification trade-off. On the one hand, for the offline case, we plan to explore automatic tools such as ParamLS [10] and F-RACE [4]. On the other hand, for the online case, we would like to investigate the use of reinforcement learning for self-adaptive tuning of the *pd* parameter. Finally, we would like to investigate the use of supervised machine learning to identify the best set of algorithms for a given problem instance [3].

References

1. Arbelaez, A., Codognet, P.: From sequential to parallel local search for SAT. In: Middendorf, M., Blum, C. (eds.) EvoCOP 2013. LNCS, vol. 7832, pp. 157–168. Springer, Heidelberg (2013). https://doi.org/10.1007/978-3-642-37198-1_14
2. Arbelaez, A., Hamadi, Y.: Improving parallel local search for SAT. In: Coello, C.A.C. (ed.) LION 2011. LNCS, vol. 6683, pp. 46–60. Springer, Heidelberg (2011). https://doi.org/10.1007/978-3-642-25566-3_4
3. Arbelaez, A., Hamadi, Y., Sebag, M.: Continuous search in constraint programming. In: Hamadi, Y., Monfroy, E., Saubion, F. (eds.) Autonomous Search, pp. 219–243. Springer, Heidelberg (2011). https://doi.org/10.1007/978-3-642-21434-9_9
4. Birattari, M., Stützle, T., Paquete, L., Varrentrapp, K.: A racing algorithm for configuring metaheuristics. In: GECCO, pp. 11–18 (2002)
5. Cai, S., Luo, C., Lin, J., Su, K.: New local search methods for partial MaxSAT. Artif. Intell. **240**, 1–18 (2016)
6. Glover, F.: Tabu search for nonlinear and parametric optimization (with links to genetic algorithms). Discrete Appl. Math. **49**(1–3), 231–255 (1994)
7. Hamadi, Y., Jabbour, S., Sais, L.: ManySAT: a parallel SAT solver. JSAT **6**(4), 245–262 (2009)
8. Hoos, H.H.: An adaptive noise mechanism for WalkSAT. In: AAAI/IAAI, pp. 655–660 (2002)
9. Hutter, F., Hoos, H.H., Leyton-Brown, K.: Tradeoffs in the empirical evaluation of competing algorithm designs. Ann. Math. Artif. Intell. **60**(1–2), 65–89 (2010)
10. Hutter, F., Hoos, H.H., Leyton-Brown, K., Stützle, T.: ParamILS: an automatic algorithm configuration framework. J. Artif. Intell. Res. **36**, 267–306 (2009)
11. Li, C.M., Wei, W., Zhang, H.: Combining adaptive noise and look-ahead in local search for SAT. In: Marques-Silva, J., Sakallah, K.A. (eds.) SAT 2007. LNCS, vol. 4501, pp. 121–133. Springer, Heidelberg (2007). https://doi.org/10.1007/978-3-540-72788-0_15
12. Luo, C., Cai, S., Su, K., Huang, W.: CCEHC: an efficient local search algorithm for weighted partial maximum satisfiability. Artif. Intell. **243**, 26–44 (2017)

13. Luo, C., Cai, S., Wu, W., Jie, Z., Su, K.: CCLS: an efficient local search algorithm for weighted maximum satisfiability. IEEE Trans. Comput. **64**(7), 1830–1843 (2015)
14. Martins, R., Manquinho, V.M., Lynce, I.: An overview of parallel SAT solving. Constraints **17**(3), 304–347 (2012)
15. Roli, A.: Criticality and parallelism in structured SAT instances. In: Van Hentenryck, P. (ed.) CP 2002. LNCS, vol. 2470, pp. 714–719. Springer, Heidelberg (2002). https://doi.org/10.1007/3-540-46135-3_51
16. Roli, A., Blesa, M.J., Blum, C.: Random walk and parallelism in local search. In: Metaheuristic International Conference (MIC 2005), Vienna, Austria (2005)
17. Selman, B., Kautz, H.A., Cohen, B.: Noise strategies for improving local search. In: AAAI, pp. 337–343 (1994)
18. Selman, B., Levesque, H.J., Mitchell, D.G.: A new method for solving hard satisfiability problems. In: AAAI 1996, pp. 440–446 (1996)
19. Strickland, D.M., Barnes, E.R., Sokol, J.S.: Optimal protein structure alignment using maximum cliques. Oper. Res. **53**(3), 389–402 (2005)
20. Thornton, J., Pham, D.N., Bain, S., Ferreira Jr., V.: Additive versus multiplicative clause weighting for SAT. In: AAAI. pp. 191–196 (2004)
21. Tompkins, D.A.D., Hoos, H.H.: UBCSAT: an implementation and experimentation environment for SLS algorithms for SAT and MAX-SAT. In: Hoos, H.H., Mitchell, D.G. (eds.) SAT 2004. LNCS, vol. 3542, pp. 306–320. Springer, Heidelberg (2005). https://doi.org/10.1007/11527695_24
22. Vasquez, M., Hao, J.: A "logic-constrained" knapsack formulation and a tabu algorithm for the daily photograph scheduling of an earth observation satellite. Comput. Optim. Appl. **20**(2), 137–157 (2001)
23. Zhang, L., Bacchus, F.: MAXSAT heuristics for cost optimal planning. In: AAAI 2012 (2012)
24. Zhang, W., Rangan, A., Looks, M.: Backbone guided local search for maximum satisfiability. In: IJCAI 2003, pp. 1179–1186 (2003)

Dynamic Compartmental Models for Large Multi-objective Landscapes and Performance Estimation

Hugo Monzón[1,4(✉)] [ID], Hernán Aguirre[1,4(✉)] [ID], Sébastien Verel[2,5] [ID],
Arnaud Liefooghe[3,5] [ID], Bilel Derbel[3,5] [ID], and Kiyoshi Tanaka[1,4]

[1] Faculty of Engineering, Shinshu University, Nagano, Japan
hugo91@gmail.com, {ahernan,ktanaka}@shinshu-u.ac.jp
[2] Université du Littoral Côte d'Opale, Calais, France
verel@lisic.univ-littoral.fr
[3] Univ. Lille, CNRS, CRIStAL, Inria Lille – Nord Europe, Lille, France
{arnaud.liefooghe,bilel.derbel}@univ-lille.fr
[4] MODŌ - International Associated Laboratory, Nagano, Japan
[5] MODŌ - International Associated Laboratory, Lille, France

Abstract. Dynamic Compartmental Models are linear models inspired by epidemiology models to study Multi- and Many-Objective Evolutionary Algorithms dynamics. So far they have been tested on small MNK-Landscapes problems with 20 variables and used as a tool for algorithm analysis, algorithm comparison, and algorithm configuration assuming that the Pareto optimal set is known. In this paper, we introduce a new set of features based only on when non-dominated solutions are found in the population, relaxing the assumption that the Pareto optimal set is known in order to use Dynamic Compartment Models on larger problems. We also propose an auxiliary model to estimate the hypervolume from the features of population dynamics that measures the changes of new non-dominated solutions in the population. The new features are tested by studying the population changes on the Adaptive ϵ-Sampling ϵ-Hood while solving 30 instances of a 3 objective, 100 variables MNK-landscape problem. We also discuss the behavior of the auxiliary model and the quality of its hypervolume estimations.

Keywords: Compartmental models · Modeling · Multi-objective optimization · Population dynamics · Hypervolume estimation

1 Introduction

Dynamic Compartmental Models (DCMs) [8,9] are linear compartmental models that track population dynamics in Multi- and Many-Objective Evolutionary Algorithm (MOEAs). They are based on epidemiology models, mainly the SIR model [4]. In the SIR model, a population of individuals is broken in groups assigned each to a compartment of the model in accordance to their health

© Springer Nature Switzerland AG 2020
L. Paquete and C. Zarges (Eds.): EvoCOP 2020, LNCS 12102, pp. 99–113, 2020.
https://doi.org/10.1007/978-3-030-43680-3_7

status, which changes as time progresses. This is captured by the model equations and parameters. Similarly, the goal of a DCM is to capture the changes of the population focusing on the dominance relationship between individuals. Each group or compartment represents how many are in a particular state of domination. The interaction between compartments and the rates of interaction is captured by the equations that defined the model and its parameters. How membership to a compartment is defined represents a feature on the population, and different sets of features allow to explore the same algorithm, problem, and configuration from other perspectives.

Two and three compartments DCMs have been successfully used to study and explain in detail the dynamics of multi-objective evolutionary algorithms. DCMs do not provide a direct estimation of performance of an algorithm expressed in terms of well known performance estimators such as hypervolume, generational distance, inverse generational distance, and others. To associate dynamics to performance, DCMs require that at least one of the compartments relates to some rate of improvement of the algorithm, from which a known performance metric can be correlated or estimated. Previous works using DCMs have focused on problems where the Pareto optimal set is known and have therefore used features associated to rates of improvement of the algorithm that require knowledge of whether a solution is Pareto optimal or not. Although DCMs on these problems have served to gain knowledge about the working principles of multi- and many-objective evolutionary algorithms, in order to use DCMs on real world scenarios, where the set of Pareto optimal solutions is unknown, new sets of features and ways to estimate measures of performance from features of population dynamics are required.

From this standpoint, in this work, we introduce new features focusing on when non-dominated solutions appear in the population to define the compartments of the model. In particular, we define a three compartments DCM where the population is divided into (1) new non-dominated solutions, (2) non-dominated but not new solutions, and (3) dominated solutions. The goal of these features is to keep track of how many new solutions appear in each generation, which serves to estimate the rate of progress of the algorithm. These features are useful whether the problem is enumerable or not. In addition, we propose an additional model to estimate a performance metric, the hypervolume achieved by an algorithm, from the features of population dynamics, i.e. new non-dominated solutions. An effective way to estimate performance from features of dynamics opens new venues to apply DCMs beyond algorithm analysis and understanding.

The paper is organized as follows. Section 2 describes in more detail the DCMs, the proposed new feature set and how to relate them to performance. Section 3 covers the experimental results to test the new features with the DCMs and the HV model. Section 4 concentrates on the proposed model for estimation of performance, and analyzes it on more configurations. Finally, in Sect. 5 we resume the work done and propose some future directions to expand it.

2   Methodology

2.1   Dynamic Compartmental Models for Multi-objective Evolutionary Algorithms

Dynamic Compartmental Models (DCM) are mathematical models that simulate how individuals in different compartments in a population interact and affect the instantaneous composition of the compartments. Here, the assumptions are that the population can be divided into compartments and that every individual in the same compartment has the same characteristics. The rates of interaction between compartments are known as the parameters of the model.

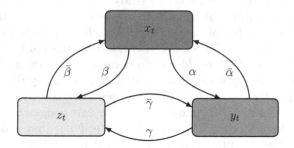

Fig. 1. A Three compartment DCM

Linear compartmental models of up to three compartments have been used to study the population dynamics of evolutionary multi-objective algorithms using the Pareto dominance status of the individuals as criteria to define the compartments. Figure 1 illustrates a three compartments DCM, which can be described by the following equations,

$$
\begin{cases}
x_{t+1} &= (1 - (\alpha + \beta))x_t + \bar{\alpha}y_t + \bar{\beta}z_t \\
y_{t+1} &= \alpha x_t + (1 - (\bar{\alpha} + \gamma))y_t + \bar{\gamma}z_t \\
z_{t+1} &= \beta x_t + \gamma y_t + (1 - (\bar{\beta} + \bar{\gamma}))z_t \\
P &= x_t + y_t + z_t,
\end{cases}
\tag{1}
$$

where x_t, y_t and z_t are variables associated to the number of individuals in the compartments at time (generation) t and α, β, γ, $\bar{\alpha}$, $\bar{\beta}$, and $\bar{\gamma}$ are the interaction rates between compartments.

Each compartment size at time $t + 1$ depends on its size and the size of all other compartments at time t modified by some constant, i.e a parameter of the model. From the system of Eq. (1) and its graphical representation on Fig. 1, we see that any change in one compartment will be distributed into the other ones, therefore the total number of individuals remains constant. This models the dynamics of an evolutionary algorithm with a fixed population size throughout the generations. It is important to note that the model tracks changes between compartments, not specific individuals.

The parameter values of the model are estimated (learned) from the data generated by the algorithm which dynamics we want to capture. The selected algorithm is run tracking on each generation the features (compartments sizes) we choose for our model. The output data relevant to these features is used to fit the model's parameters. Thus, the parameters of the model are linked to a particular algorithm set with a given configuration on a problem instance or a subclass of problems. If the algorithm, its configuration or the problem changes, parameters naturally will change too.

DCMs have been successfully used to study and explain in detail the dynamics of multi- and many-objective evolutionary algorithms, gaining knowledge about the working principles of the various approaches to design these algorithms [8,9]. DCMs were used, for example, to study how multi- and many-objective evolutionary algorithms are able to continue discovering Pareto optimal solutions once their population is full of them in order to achieve a high resolution of the Pareto optimal set (POS). To answer this question the study was conducted on problems where the POS could be enumerated [9], defining the three compartments so that the population was divided into (1) newly discovered Pareto optimal solutions, (2) non-dominated but not new Pareto optimal solutions, and (3) dominated solutions. The union of the first two compartments is the set of non-dominated solutions in the population. Verifying that a non-dominated solution is also a Pareto optimal solution and that it has been seen by the algorithm for the first time in the current generation allows dividing non-dominated solutions into the two first compartments mentioned above. Of course, this can be done if and only if the POS is known.

As mentioned before, DCMs directly do not estimate the performance of an algorithm in terms of well known and commonly used estimators such as hypervolume, generational distance, inverse generational distance, and others. However, to associate the dynamics to performance is possible to have a feature set where at least one of them can carry information about the rate of improvement of the algorithm, which then can be correlated or used to estimate a more common performance metric. In [8,9] the first compartment referred above, i.e. the number of new Pareto optimal solutions in the population, provides the rate of discovery of Pareto optimal solutions and gives a rate of improvement of the algorithm. Thus, in these works, this feature was correlated to performance. Namely, it was shown that the accumulation of newly discovered Pareto optimal solutions is highly correlated to the hypervolume. In other words, it is possible to look at this feature to decide with high confidence what algorithm (or algorithm configuration) is better than others.

DCMs can also be used to predict future behavior and performance of the algorithm. That is, running the DCM for additional generations for which the actual algorithm has not been yet run can be estimated with high confidence, for example, whether increasing the fitness evaluation budget for a giving algorithm may translate into improved performance. This is quite relevant to application domains where fitness is computationally expensive, such as simulation-based optimization.

Another important potential use of DCMs is for algorithm configuration and algorithm selection [9]. For example, let us assume we want to configure population size for a given budget of fitness evaluations. A common approach is to run the algorithm several times, each time with a different combination of population size and number of generations. Another alternative is to run the algorithms in a sample of configurations, learn DCMs for each one of them, and infer new models for intermediate configurations by interpolation of the models' parameters.

Initial explorations of the application of DCMs are promising. However, the above studies have been done on small landscapes and using a feature that requires knowledge of the Pareto optimal set. In order to use DCMs in real-world scenarios, dynamics should relate to performance using features that correlate to a rate of improvement of the algorithm but do not require to know whether a solution is optimal or not.

In the next sections, we introduce a set of features that can be used for such purpose together with a method to estimate performance from one of them.

2.2 The NDNew-NDOld-DOM Feature Set

To explore DCMs in large problems, we define three compartments so that the population is divided into (1) new non-dominated solutions, (2) non-dominated but not new solutions, and (3) dominated solutions. These compartments or features are called for short *Non-Dominated New*, *Non-Dominated Old*, and *Dominated*. A solution is counted as Non-Dominated New at generation t only if it is a non-dominated solution in the population but did not appear in any previous generation from 0 to $t-1$. A solution is counted as Non-Dominated Old at generation t if it is a non-dominated solution and has also appeared in a previous generation. A solution is counted as Dominated at generation t if it is a dominated solution in the population. A more compact explanation can be seen in Table 1. While this set of features does not offer directly a way to measure performance, it still gives an idea of the progress of the search, since we expect to see the number of New Non-Dominated solutions to go down when the algorithm is converging.

It is important to mention that when we count a Non-Dominated solution at generation t, it is non-dominated relative to the population at that generation. It may be that at a future generation that solution becomes dominated. We could maintain an updated list of solutions non-dominated so far and check non-dominated solutions in the current population against it before counting it. However, this could add substantial computational overhead and it is not clear whether this could add any extra value to the feature. As it is defined now, it still serves the purpose of showing us from the algorithm perspective if the search is still moving, i.e. it has not stagnated.

Table 1. Proposed features. $\mathcal{F}_1$: first front containing all the non-dominated solutions. t: current generation. P: whole population including $\mathcal{F}_1$.

Abbr.	Formula	Comment
NDNew	$\{x : x \in \mathcal{F}_1(t) \wedge x \notin \cup_{k=0}^{t-1} \mathcal{F}_1(k)\}$	New non dominated solutions
NDOld	$\{x : x \in \mathcal{F}_1(t) \wedge x \in \cup_{k=0}^{t-1} \mathcal{F}_1(k)\}$	Old non dominated solutions
DOM	$\{x : x \in P \wedge x \notin \mathcal{F}_1(t)\}$	Dominated solutions

2.3   Performance Metrics and Features

The new features of dynamics do not provide a direct measure of the algorithm performance from the model. One solution to this issue is creating an auxiliary model that takes in some of the features and an initial evaluation of a performance metric to estimate this value at any generation. In this work, we estimate the hypervolume indicator (HV) [11], more specifically the hypervolume calculated over the Non-Dominated set of all solutions in the population at generation t and previous ones. The reference point is set to $(0,0,0)$. Figure 2 illustrates the model learning process of population dynamics and performance features from some sampled configurations. We try a model of the form $HV_{t+1} = HV_t + \mu \times \textit{some feature}\ /t$.

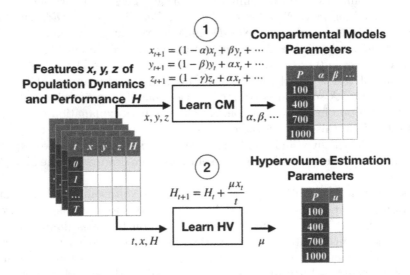

Fig. 2. Scheme of the model learning process of population dynamics and performance features from some sampled configurations.

We used Grammatical Evolution, a tool from Genetic Programming, that searches for expressions instead of programs. To evaluate which expression gives the best model, we use the mean square error between the model and our reference data, namely the number of found NDNew, NDOld and DOM solutions in generation t and the corresponding hypervolume HV for generation $t + 1$. The first step is defining a grammar than can derive in the type of expression we need, in this case, $\mu \times$ *some feature*$/t$, which we will refer to as ΔHV_t, and is presented in Fig. 3.

$$
\begin{aligned}
\langle expr \rangle &\models \langle expr \rangle \langle op \rangle \langle expr \rangle \mid \text{-1*}\langle value \rangle \text{*}\langle var \rangle \mid \langle value \rangle \text{*}\langle var \rangle \\
\langle op \rangle &\models + \mid - \mid \times \mid \hat{\ } \mid / \\
\langle var \rangle &\models \text{NDNew} \mid \text{NDOld} \mid \text{t} \\
\langle value \rangle &\models \langle cat \rangle \\
\langle cat \rangle &\models \langle int \rangle.\langle int \rangle \mid \langle int \rangle \\
\langle int \rangle &\models \langle int \rangle \langle number \rangle \mid \langle number \rangle \\
\langle number \rangle &\models [\text{0-9}]
\end{aligned}
$$

Fig. 3. BNF Grammar used to search for an expression that relates the hypervolume to the features.

This grammar can generate expressions such as $0.833 \times \text{NDNew}/\text{NDOld}$ or $-5 \times \text{NDOld} \times t$. To implement this part we used gramEvol [10] a library available in the R language. After some tries with this library the suggested expression for the model was:

$$
\text{HV}_{t+1} = \text{HV}_t + \frac{\mu \times \text{NDNew}_t}{t+1}, \tag{2}
$$

The model can be interpreted as the HV will grow on generation $t+1$ proportionally to how many New Non-Dominated solutions were found at generation t times a constant μ and inversely to the next generation number. This makes sense as the impact of finding solutions at the beginning will surely make the hypervolume value jump, while at the end we can think that these newly found solutions probably fill in gaps having very little effect.

3   Experimental Results

3.1   Test Problem and Experiment Settings

Testing these new features requires generating some data by running an MOEA with different configurations on a given problem. The MOEA we selected is the Adaptive ϵ-Sampling ϵ-Hood (AϵSϵH), a Many-objective Optimization Evolutionary Algorithm that can also handle Multi-objective problems. Its approach is Pareto dominance relaxation in the form of ϵ-dominance to determine which solutions are kept and how parents are selected for the next generation [1]. The crossover is two-point with rate $pc = 1$, bit flip mutation with rate $pm = 1/N$, the reference neighborhood size is set to 20 individuals and the ϵ-dominance function is additive ($f' = f + \epsilon$).

The chosen problem is the combinatorial multi-objective problem generator MNK-Landscapes [2]. Its parameters are the number of objectives M, number of variables N and K, a value that allows setting the ruggedness by determining the number of epistatic interactions between variables. This is, it determines how much other variables affect the fitness contribution of a given variable. In MNK-Landscapes terms, an M = 3, N = 100 and K = 5 problem is a 3 objective, 100 variables one where each variable fitness contribution will be affected by other 5 variables values defined as part of the problem. We generated 30 landscapes or sub-classes of an M = 3, N = 100 and K = 5 problem, each time the epistatic interactions are determined at random when the problem is created.

The data for the models is then generated by running AεSεH on each of the 30 landscapes, with different configurations, i.e, population sizes ranging from 3000 to 10500, with increments of 2500. On each configuration, the maximum number of Function Evaluations (FE) allowed was of 600000, which determined the maximum number of generations to run the algorithm (FE = Population Size $\times$ t_{max}). In this section, results will be shown for models that only have seen data until 400000 FE, and in the next one, we will present results with other FE limits.

Lastly, when we talk about the models' estimation we want to emphasize that for our DCMs we give only one measured value, the ones obtained from generation 0, i.e. the initial population. Here is also where we measure the first hypervolume value used to start the HV model. From there, both models use the estimation they generated for generation t to calculate the following one in $t + 1$, and so on, until the required number of generations t_{max} is met.

3.2   Fitting of the Models

The fitting process, was done with the Levenberg-Marquardt Non-Linear Least Squares algorithm [6,7] using the R language implementation [3]. The input for this process is the feature data from the algorithms and the system of equations (1), obtaining the parameters for some configurations with different population sizes, $(3000, 5500, 8000, 10500)$ and varying the FE limits, $(300000, 400000, 500000, 600000)$.

We took some considerations while doing the fitting process. Instead of using the feature data from each landscape's data, we take the average value of the features at each generation, including the HV value. This, at least for the DCM had a more significant impact on producing better estimations.

Cross-validation was also introduced, so the obtained parameters are not a product of over-fitting to our generated data and would generalize better in the presence of new and unseen data. We choose k-fold cross-validation and apply it during DCMs and HV model fitting process. In k-fold cross-validation, the dataset is split into k subsets of equal size, each subset is used only once as a test set and $k - 1$ times as part of the training set. For our data, we have 30 runs of the algorithm, corresponding each one to a different landscape. To ensure an 80/20 split between training and testing data, we select $k = 5$, a common recommendation for this method as suggested in [5]. So each fold is composed of one subset of 6 landscapes worth of test data, and the remaining 4 subsets, provide 24 landscapes worth of training data. The score obtained on each set is measured by the goodness of fit or R^2, a value between 0 and 1 that indicates how much of the variance present in the data is explained by the model.

Under cross-validation, the fitting process per configuration is done only with the data from the training set, and the resulting parameters estimation ability is measured on the test set, repeated for each fold. We report in Table 2 the average R^2 of the 5 scores obtained for the training and test datasets for all population sizes and considering only 400000 FE available. Since at the end of the process we also have 5 sets of parameters, we take the average and keep the result as the best parameters found for that configuration and number of FE.

From the table, we can see that the R^2 is overall higher than 0.64 for the features NDOld, DOM and HV for both training and testing sets, while the NDNew feature has a lower score when compared to the other features. To understand better this situation, we can refer to Figs. 4, 5, 6 and 7 that show the NDNew-NDOld-DOM feature measured values and the DCM estimation for all the considered population sizes. In these plots, we are using the obtained best parameters and creating the estimations considering all available landscapes.

Table 2. R^2 values obtaining during the model training and testing. Obtained doing k-fold cross-validation with $k = 5$. Considering only 400000 FE.

	Training R^2				Testing R^2			
Population	NDNew	NDOld	DOM	HV	NDNew	NDOld	DOM	HV
3000	0.69580	0.86617	0.84022	0.70278	0.85538	0.82754	0.66148	0.66104
5500	0.65387	0.82545	0.80737	0.66051	0.81289	0.79353	0.86231	0.86223
8000	0.64272	0.87690	0.86889	0.64110	0.85950	0.85072	0.91488	0.91099
10500	0.68936	0.87161	0.86475	0.65749	0.86807	0.86062	0.88234	0.87566

From a first look at the figures, we can see that the model estimation (red points), goes through the middle of the measured data (black points) since the fitting was done on the average values for the features. Not doing so, produced overestimations after the hump in the feature NDNew graph, which translated into a poor HV estimation as the other model depends on this value. The lower R^2 in this feature, when compared to the other two, could be attributed to the higher overall variance present as appreciated in the figure. An understandable situation, since the number of newly found non-dominated solutions, can change very quickly.

It is also interesting to notice how this simple model can adapt to different configurations, for larger population sizes the number of generations diminishes and the fitted model can keep up with the different rates of change in each case.

Now we move to the HV model results, in Fig. 8 we show the estimation against measured values for all sampled population sizes. As can be seen, the model seems to follow the change of the hypervolume until a certain point from

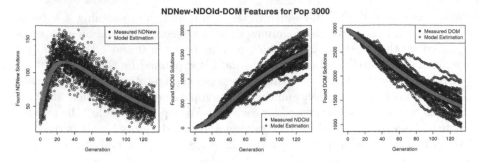

Fig. 4. Non-Dominated New, Non-Dominated Old and Dominated Solutions DCM's estimation vs Measured data for a configuration with Pop. Size 3000 and 400000 FE.

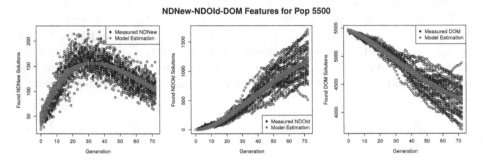

Fig. 5. Non-Dominated New, Non-Dominated Old and Dominated Solutions DCM's estimation vs Measured data for a configuration with Pop. Size 5500 and 400000 FE.

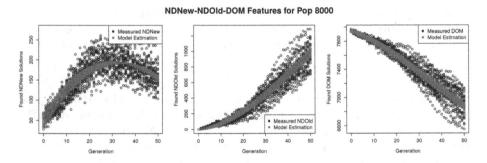

Fig. 6. Non-Dominated New, Non-Dominated Old and Dominated Solutions DCM's estimation vs Measured data for a configuration with Pop. Size 8000 and 400000 FE.

which there is a tendency to overestimate in all configurations. Looking aside from the overestimation in the last few generations, it seems to follow the overall tendency of the hypervolume, which growth seems correlated to the NDNew feature.

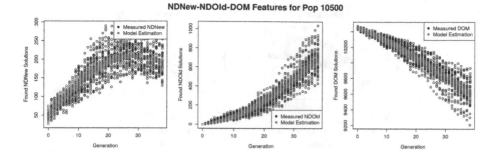

Fig. 7. Non-Dominated New, Non-Dominated Old and Dominated Solutions DCM's estimation vs Measured data for a configuration with Pop. Size 10500 and 400000 FE.

4 Discussion

In the last section, we discussed how to create and fit both, the Dynamic Compartmental Model and the HV model. Here we want to explore how well they can to follow the trend in the performance data. We will look at the estimated accumulated HV at the end of a run for different population sizes and the maximum number of Function Evaluations allowed. As mentioned before, the hypervolume at each generation is calculated over all non-dominated solutions found until that generation t, therefore we refer to it as accumulated HV. In Figs. 9, 10, 11 and 12 we have two box plots per population size, the one in red represents the measured data (M) for all 30 landscapes while the one in blue is the HV model estimation (E) for the same 30 landscapes with each figure showing results considering 300000, 400000, 500000 and 600000 FE.

Looking at the big picture, we notice that for every variation of the FE the measured data indicates a downward trend. That is, even though we keep adding more FE so the largest population sizes could benefit with more time to get a better convergence, this does not translate into a better overall final HV. Thus, in this particular problem, it seems that a population size of 3000 is enough to ensure a good final hypervolume. If we look now at the model estimation, we notice a clear overestimation for all population sizes, though it maintains the ordering, replicating the trend seen in the data. This is particularly important if we want to use the models for any analysis and to distinguish which configurations perform better than others.

If we focus on the plot of the NDNew feature on Figs. 4, 5, 6 and 7, we see that our DCM learned the mean of the data and this still produces overestimation as can be checked in the plot against the measured data in Fig. 8. In fact, for all the variations in FE, the DCM keeps going through the mean and still ends in an overestimation when used by our current HV model.

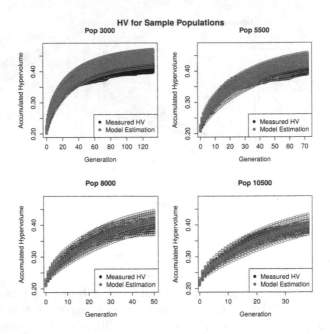

Fig. 8. HV model estimation vs Measured data for sampled configurations with Pop. Size [3000,5500,8000,10500] and 400000 FE.

From the formulation, it seems that we are on the right track and there is a connection between newly discovered solutions appearance rate and the growth of the HV, but our parameter μ, or even the current generation number do not seem enough to keep this estimation closer to the measured values. In particular, it is important to tell the model that the weight of newly found solutions varies depending on what stage we are in the algorithm's run. In the beginning, it is not strange to see the HV grow quicker with each newly found solution, while by the end we expect these solutions to fill gaps on a set of non-dominated solutions that form a good approximation set for this problem.

Even with the current formulation, it is still interesting to see how a simple feature such as the number of newly discovered non-dominated solutions per generation can carry enough information that can be translated into what kind of trend we can expect of a performance metric such as the hypervolume. More so if we remember that for all the estimations done with the models, they only start with one piece of measured data, and from there is purely the captured dynamics and behavior of the algorithm that guides the process.

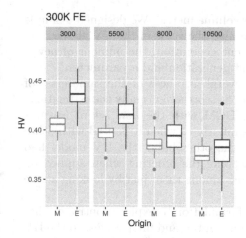

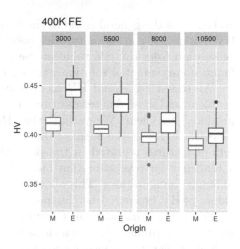

Fig. 9. Comparison between the final HV for all landscapes on 300000 FE.

Fig. 10. Comparison between the final HV for all landscapes on 400000 FE.

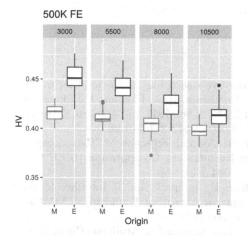

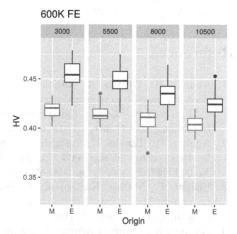

Fig. 11. Comparison between the final HV for all landscapes on 500000 FE.

Fig. 12. Comparison between the final HV for all landscapes on 600000 FE.

5   Conclusions and Future Work

In this work we proposed a new set of features that allows Dynamic Compartmental Models to be used on larger multi-objective problems where the Pareto optimal set is not known or cannot be obtained through enumeration, removing the assumption that the Pareto optimal set is known made by previously proposed feature sets. Parting from the knowledge that features that capture the rate of improvement of an algorithm can be correlated to a performance metric, we presented and tested a possible auxiliary model that can estimate and capture the general trend of the hypervolume metric. We designed a simple HV model that can estimate, with good results, the value of the HV at the next generation from the HV value at the current generation, the number of newly found non-dominated solutions, the current generation and a parameter.

We tested DCM and HV models on several instances of the same class of problem, and showed in terms of goodness of fit score and visually the estimations produced by them. We verified that the DCM with the new set of features successfully learns the mean of the data, similarly to when it is used with other sets of features reported in the past. On the other hand, the HV model had a tendency for overestimation but still keeping the ordering when applied on different configurations. This allows selecting among them just by looking at the values of HV estimated by the model.

For future work, we want to revise the formulation to explore control mechanisms to discriminate between the algorithm's initial and final stages, so the HV estimation can be smoother and closer to the measured values. We also plan to introduce interpolation of the parameters and use it for selecting configurations, exploiting the relationship between a set of parameters and the configuration and algorithm from which it was obtained.

References

1. Aguirre, H., Oyama, A., Tanaka, K.: Adaptive ϵ-sampling and ϵ-hood for evolutionary many-objective optimization. In: Purshouse, R.C., Fleming, P.J., Fonseca, C.M., Greco, S., Shaw, J. (eds.) EMO 2013. LNCS, vol. 7811, pp. 322–336. Springer, Heidelberg (2013). https://doi.org/10.1007/978-3-642-37140-0_26
2. Aguirre, H., Tanaka, K.: Insights on properties of multiobjective MNK-landscapes. In: Proceedings of the 2004 Congress on Evolutionary Computation (IEEE Cat. No.04TH8753), vol. 1, pp. 196–203, June 2004
3. Elzhov, T., Mullen, K., Spiess, A., Bolker, B.: minpack. lm: R Interface to the Levenberg-Marquardt Nonlinear Least-Squares Algorithm Found in MINPACK, Plus Support for Bounds (ver. 1.2-0) r package (2015)
4. Godfrey, K.: Compartmental Models and Their Application. Academic Press, Cambridge (1983)
5. James, G., Witten, D., Hastie, T., Tibshirani, R.: An Introduction to Statistical Learning, vol. 112. Springer, New York (2013). https://doi.org/10.1007/978-1-4614-7138-7
6. Levenberg, K.: A method for the solution of certain non-linear problems in least squares. Q. Appl. Math. $\mathbf{2}$(2), 164–168 (1944)
7. Marquardt, D.W.: An algorithm for least-squares estimation of nonlinear parameters. J. Soc. Ind. Appl. Math. $\mathbf{11}$(2), 431–441 (1963)
8. Monzón, H., Aguirre, H., Verel, S., Liefooghe, A., Derbel, B., Tanaka, K.: Dynamic compartmental models for algorithm analysis and population size estimation. In: Proceedings of the Genetic and Evolutionary Computation Conference Companion, pp. 2044–2047. ACM (2019)

9. Monzón, H., Aguirre, H.E., Verel, S., Liefooghe, A., Derbel, B., Tanaka, K.: Closed state model for understanding the dynamics of MOEAs. In: Proceedings of the Genetic and Evolutionary Computation Conference, GECCO 2017, Berlin, Germany, 15–19 July 2017, pp. 609–616 (2017)

10. Noorian, F., de Silva, A.M., Leong, P.H.W.: gramEvol: grammatical evolution in R. J. Stat. Softw. **71**(1), 1–26 (2016). https://doi.org/10.18637/jss.v071.i01

11. Zitzler, E., Thiele, L.: Multiobjective evolutionary algorithms: a comparative case study and the strength pareto approach. IEEE Trans. Evol. Comput. **3**(4), 257–271 (1999)

Fitness Landscape Analysis of Automated Machine Learning Search Spaces

Cristiano G. Pimenta[1(✉)], Alex G. C. de Sá[1], Gabriela Ochoa[2],
and Gisele L. Pappa[1]

[1] Computer Science Department, Universidade Federal de Minas Gerais,
Belo Horizonte, Brazil
{cgpimenta,alexgcsa,glpappa}@dcc.ufmg.br
[2] University of Stirling, Stirling, UK
gabriela.ochoa@stir.ac.uk

Abstract. The field of Automated Machine Learning (AutoML) has as its main goal to automate the process of creating complete Machine Learning (ML) pipelines to any dataset without requiring deep user expertise in ML. Several AutoML methods have been proposed so far, but there is not a single one that really stands out. Furthermore, there is a lack of studies on the characteristics of the fitness landscape of AutoML search spaces. Such analysis may help to understand the performance of different optimization methods for AutoML and how to improve them. This paper adapts classic fitness landscape analysis measures to the context of AutoML. This is a challenging task, as AutoML search spaces include discrete, continuous, categorical and conditional hyperparameters. We propose an ML pipeline representation, a neighborhood definition and a distance metric between pipelines, and use them in the evaluation of the fitness distance correlation (FDC) and the neutrality ratio for a given AutoML search space. Results of FDC are counter-intuitive and require a more in-depth analysis of a range of search spaces. Results of neutrality, in turn, show a strong positive correlation between the mean neutrality ratio and the fitness value.

Keywords: Fitness landscape analysis · Automated Machine Learning · Fitness distance correlation · Neutrality

1 Introduction

The recent hype on machine learning (ML) and its application to a wide range of problems that are close to the general public has increased the interest in the area and, consequently, the number of people using ML to solve a wide range of problems [24]. However, the performance of an ML solution to a specific learning problem depends heavily on the choice of data preprocessing and learning algorithms, as well as on their hyperparameters. Although the choice of an ML solution can be manually made, this is a hard and not effective process.

© Springer Nature Switzerland AG 2020
L. Paquete and C. Zarges (Eds.): EvoCOP 2020, LNCS 12102, pp. 114–130, 2020.
https://doi.org/10.1007/978-3-030-43680-3_8

Considering the number of ML algorithms, combinations among them and their associated hyperparameters, the number of choices can grow exponentially. Furthermore, manual tuning requires an inherent expertise in the choice of methods and the possible values of their hyperparameters [24]. Hence, it is a challenging task for people who have little knowledge in ML.

The area of *Automated Machine Learning* (AutoML) has emerged as a solution to the aforementioned issue, becoming very popular over the past decades [8]. Its main objective is to automate the process of recommending or creating complete machine learning pipelines to any dataset without requiring deep user knowledge on the learning task itself [8]. A machine learning pipeline is a sequence of tasks to follow when performing data analysis in a specific dataset. It can include preprocessing steps (e.g., data cleaning, data discretization and feature selection [23]), a machine learning model (such as a classifier or a regressor), and postprocessing steps that may help to combine the results of several ML models (for instance, a voting method [23]).

Several optimization methods have been proposed to solve the problem of automatically generating ML pipelines, including those based on Bayesian optimization (e.g., Auto-WEKA and AutoSKLearn), evolutionary search (e.g., RECIPE and TPOT), multi-fidelity optimization (e.g., Hyperband) and hierarchical planing (e.g., ML-Plan) [8,24]. However, there is not a single one that seems to outperform all the others and, in most cases, very similar results are obtained by different methods.

AutoML methods based on optimization techniques rely on two main components: a search space and an optimization method. The search space comprises the main building blocks (e.g., the preprocessing methods, the learning models, the postprocessing approaches and their associated hyperparameters) from previously designed ML pipelines. The optimization method is responsible for finding the best combinations of ML components to build the most effective pipelines according to a quality metric to a given dataset.

There is still very little knowledge on how the characteristics of the search space impact AutoML methods. These search spaces are difficult to analyze, as they include discrete, continuous, categorical and conditional variables [8]. A better understanding of AutoML search spaces can help to explain the performance of existing algorithms and lead to the development of new ones, designed to explore the peculiarities of these spaces [14].

One way to analyze the characteristics of the search spaces is through fitness landscape analysis (FLA) [19]. The fitness landscape of a problem is given by the values of fitness obtained by all possible solutions present in the search space. The idea of FLA methods is to gain a better understanding of algorithm performance on a related set of problem instances, creating an intuitive understanding of how a heuristic algorithm explores the fitness landscape. However, as AutoML search spaces contain mixed types of variables, performing FLA in this case is more challenging because the notion of neighborhood or distance function needed by FLA metrics is not straightforward.

Fitness landscape analysis of algorithm configuration and machine learning pipeline generation is still in its early stages [7,15]. In this paper, we propose a way of measuring the distance between machine learning pipelines and adapt typical FLA metrics to the complex search spaces of AutoML. We then adapt the fitness distance correlation (FDC) and the neutrality ratio metrics for AutoML search spaces. Results of FDC are initially counter-intuitive and require a more in-depth analysis of a range of search spaces. Results of neutrality, in turn, show a strong positive correlation between the mean neutrality ratio and the fitness value. A next step is to investigate whether this is beneficial or detrimental to different search methods.

2   Problem Definition

Before defining the fitness landscape of an AutoML problem, we formally define the problem itself. AutoML can be cast as the *Combined Algorithm Selection and Hyperparameter optimization* (CASH) problem [6,20]. Given a set $\mathcal{A} = \{A^{(1)}, A^{(2)}, \ldots, A^{(k)}\}$ of learning algorithms, where each algorithm $A^{(j)}$ has a hyperparameter space $\Lambda^{(j)}$, the CASH problem is defined in Eq. 1. In its original formulation [20], CASH is defined as a minimization problem. Here we cast it as a maximization problem, replacing the loss function with a gain function.

$$A^*_{\lambda^*} \in \operatorname*{argmax}_{A^{(j)} \in \mathcal{A}, \lambda \in \Lambda^{(j)}} \frac{1}{k} \sum_{i=1}^{k} \mathcal{G}(A^{(j)}_{\lambda}, \mathcal{D}^{(i)}_{\text{train}}, \mathcal{D}^{(i)}_{\text{valid}}), \tag{1}$$

where $\mathcal{G}(A^{(j)}_{\lambda}, \mathcal{D}^{(i)}_{\text{train}}, \mathcal{D}^{(i)}_{\text{valid}})$ is the gain achieved when a learning algorithm A, with hyperparameters λ, is trained and validated on disjoint training and validation sets $\mathcal{D}^{(i)}_{\text{train}}$ and $\mathcal{D}^{(i)}_{\text{valid}}$, respectively, on each partition $1 \leq i \leq k$ of a k-fold cross-validation procedure. The main idea of this paper is to analyze the characteristics of the search space of algorithms $\mathcal{A}$ and the hyperparameters $\lambda^{(j)}$ of each $A_j \in \mathcal{A}$.

3   AutoML Fitness Landscape

Stadler [19] defines a fitness landscape as having three components: (i) a set X of configurations; (ii) a notion $\mathcal{X}$ of neighborhood or distance on X, and (iii) a fitness function $f : X \to \mathbb{R}$. The set X of configurations and the neighborhood definition $\mathcal{X}$ define the configuration space of the problem. Depending on $\mathcal{X}$, one fitness function can be associated with several different fitness landscapes [14]. The next sections discuss these three components in the context of AutoML.

3.1   Configurations

The first component of a fitness landscape is a set X of configurations. In the case of the AutoML problem tackled in this paper, X corresponds to all valid *classification machine learning pipelines* that can be generated to solve a given problem. An ML pipeline can be defined as the sequence of algorithms that transform a feature vector $\vec{x} \in \mathbb{X}^d$ (with d dimensions) into a target vector $\vec{y} \in \mathbb{Y}$, which contains discrete values (i.e., class labels) for classification problems [24].

Machine Learning Pipelines: A typical machine learning pipeline is composed of preprocessing steps (e.g., data cleaning and feature selection), a machine learning modeling step (e.g., a classification or regression algorithm), and some postprocessing steps that may help to combine the results of several models [23].

For the sake of simplicity, in this first analysis we consider only classification pipelines composed of up to three preprocessing algorithms and one classifier, without any postprocessing steps. The pipelines are generated by creating derivation trees from a proposed context-free grammar (CFG). One of the benefits of using CFGs [18] to represent the AutoML search spaces is that they organize prior knowledge (from specialists) about the problem, properly guiding the optimization process. In addition, the grammar also gives flexibility in the definition of the search space, as the grammar rules can be modified anytime. Finally, the grammar can introduce semantics along with its syntax, possibly allowing the evaluation of the complexity of the search space.

The grammar defines the order of the preprocessing algorithms and guarantees a classification algorithm is always present in a pipeline. The search space the grammar defines is composed of 18 preprocessing and 23 classification algorithms. Given these algorithms an their associated hyperparameters, the grammar contains 148 terminal symbols and 128 non-terminal symbols and production rules, generating a search space with an estimated size of $7.88\mathrm{e}9(feat - 1)^2 + 1.62\mathrm{e}18(feat - 1) + 1.05\mathrm{e}13$ pipeline configurations, where $feat$ is the number of features of the dataset[1]. The complete grammar and other supplementary material are available online[2].

An example of a pipeline, which is a derivation tree from the grammar, is shown in Fig. 1, where algorithm names correspond to tree nodes with sharp edges. Hyperparameter names are represented as rounded rectangles with dashed lines, whereas their values correspond to the ellipses. In this paper, pipelines are initialized at random by uniformly choosing production rules from the grammar.

3.2   Neighborhood and Distance Between Pipelines

The second component of a fitness landscape is a notion $\mathcal{X}$ of neighborhood or a distance metric between the elements of the set X of configurations, described in Sect. 3.1. Considering the complexity of the AutoML search space, which contains categorical, discrete, continuous and conditional parameters, and the lack of literature on the analysis of such spaces, we propose a simple neighborhood

---

[1] When determining the search space size, for continuous hyperparameters, we simplify and always consider 100 values, regardless of the size of the interval.

[2] https://cgpimenta.github.io/EvoCOP2020_CGPimenta/.

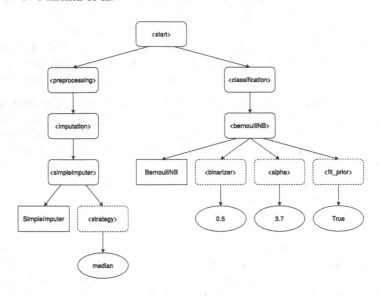

Fig. 1. Tree representation of a machine learning pipeline.

definition and a distance metric between our tree-based pipelines, adapted from the metric proposed by Ekárt & Németh for genetic programs [5].

Neighborhood Definition: We defined the neighborhood $N(s)$ of machine learning pipeline s as the set that contains all trees that result from the application of a mutation operator on a random node of the tree (i.e., the selected node is replaced by another component generated by its parent node on the tree). Figure 2 shows two neighbors of the pipeline from Fig. 1. Grey subtrees indicate where the mutation operation was employed. Considering that changing an algorithm by another can have a much bigger impact than changing the value of a hyperparameter, we define the probability $p(x)$ of choosing a node x from tree T as the mutation point as a function that increases with the distance from the root. The exceptions are the terminal symbols (leaves of the tree), which are only changed when their parent node is selected. In order to do that, we give a weight $w(x)$ to each node that is directly proportional to the probability of choosing it, as defined in Eq. 2.

In this way, the root tree (`<start>` symbol) has weight 1, and the weight increases according to the level of the node in the tree. In our case, the preprocessing symbol has weight 2, the classification and preprocessing subgroups weight 3. The non-terminals representing algorithm names have weight 4 and those representing hyperparameter names, weight 5. $p(x)$ is then given by Eq. 3.

$$w(x) = \begin{cases} \text{tree level} & \text{if non-terminal symbol} \\ 0 & \text{if terminal symbol} \end{cases} \quad (2)$$

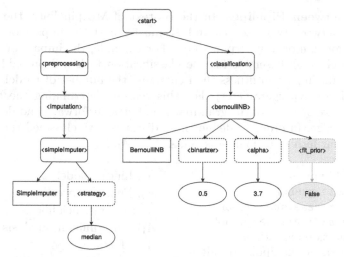

(a) Value of the `fit_prior` hyperparameter changed from
True to **False**.

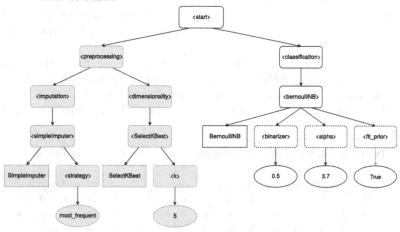

(b) Whole preprocessing subtree mutated.

Fig. 2. Two neighbors of the pipeline from Fig. 1. Grey subtrees indicate mutation
points.

$$p(x) = \frac{w(x)}{\sum_{x \in T} w(x)} \qquad (3)$$

Garciarena *et al.* [7] proposed a similar definition for the neighborhood of
an ML pipeline, where they changed a randomly chosen algorithm or hyperparameter by another feasible value. The neighborhood we proposed here can be
considered as an extension of their approach, and is based on the typical mutation operator used in grammar-based genetic programming [11]. For example,
the proposed neighborhood can change all preprocessing steps of a pipeline at
once, something the original approach does not allow.

Distance Between Pipelines: In the context of ML pipelines, the distance $dist(T_x, T_y)$ between two trees T_x and T_y must reflect the impact of changing either an algorithm or a hyperparameter. For example, the impact of changing a linear model by an ensemble for the classification task will probably have a more significant impact in fitness than changing the number of models used by the ensemble (a hyperparameter). For this reason, determining the distances between any two symbols of the grammar is not straightforward and depends on expert knowledge. In order to define these distances, we classified the symbols of the grammar into 17 disjoint sets $A_0, A_1, \ldots, A_{16}$:

- A_0: *NULL* symbol
- A_1: `<start>` symbol
- A_2: `<preprocessing>` symbol
- A_3: `<classification>` symbol
- A_4: Imputation algorithms
- A_5: Data range manipulation algorithms
- A_6: Dimensionality manipulation algorithms
- A_7: Naïve Bayes

- A_8: Linear models
- A_9: Neural networks
- A_{10}: Nearest neighbors
- A_{11}: Discriminant analysis
- A_{12}: Trees
- A_{13}: Ensembles
- A_{14}: Discrete hyperparamenters
- A_{15}: Continuous hyperparameters
- A_{16}: Categorical hyperparameters

A list of the algorithms in each set is available online (See footnote 2). The set A_0 is reserved to a special *NULL* symbol, used to treat cases in which a node in a tree does not have a corresponding node in the other. For $i, j \in \{0, 1, \ldots, 16\}$, the distance $d(x, y)$ between two symbols $x \in A_i$ and $y \in A_j$ is defined as a constant

Table 1. Distances $d(x, y)$ between symbols w.r.t. their partitions.

	A_0	A_1	A_2	A_3	A_4	A_5	A_6	A_7	A_8	A_9	A_{10}	A_{11}	A_{12}	A_{13}	A_{14}	A_{15}	A_{16}
A_0	1	0	8	0	4	4	4	0	0	0	0	0	0	0	0	0	0
A_1	0	1	0	0	0	0	0	0	0	0	0	0	0	0	0	0	0
A_2	8	0	1	0	0	0	0	0	0	0	0	0	0	0	0	0	0
A_3	0	0	0	1	0	0	0	0	0	0	0	0	0	0	0	0	0
A_4	4	0	0	0	1	0	0	0	0	0	0	0	0	0	0	0	0
A_5	4	0	0	0	0	1	0	0	0	0	0	0	0	0	0	0	0
A_6	4	0	0	0	0	0	1	2	2	2	2	2	2	2	0	0	0
A_7	0	0	0	0	0	0	2	1	2	2	2	2	2	2	0	0	0
A_8	0	0	0	0	0	0	2	2	1	2	2	2	2	2	0	0	0
A_9	0	0	0	0	0	0	2	2	2	1	2	2	2	2	0	0	0
A_{10}	0	0	0	0	0	0	2	2	2	2	1	2	2	2	0	0	0
A_{11}	0	0	0	0	0	0	2	2	2	2	2	1	2	2	0	0	0
A_{12}	0	0	0	0	0	0	2	2	2	2	2	2	1	2	0	0	0
A_{13}	0	0	0	0	0	0	2	2	2	2	2	2	2	1	0	0	0
A_{14}	0	0	0	0	0	0	0	0	0	0	0	0	0	0	0.5	0.5	0.5
A_{15}	0	0	0	0	0	0	0	0	0	0	0	0	0	0	0.5	0.5	0.5
A_{16}	0	0	0	0	0	0	0	0	0	0	0	0	0	0	0.5	0.5	0.5

that depends on the class a symbol belongs to. If x and y have the same label, $d(x, y) = 0$. Table 1 shows the values of the constants used in this work.

Note that to make this analysis possible, the grammar defined here limits the pipelines to have at most three preprocessing algorithms and exactly one classifier. Given this restriction, we represent a tree T_i with root r_i as $T_i = r_i(c_1^{(i)}, c_2^{(i)}, \ldots, c_m^{(i)})$, where the root has m children nodes, denoted by $c_j^{(i)}$, $j \in \{1, 2, \ldots, m\}$. Each node is represented by its label (i.e., its name) and can be considered as the root of a subtree. Let us consider T_x as the pipeline in Fig. 1. r_x corresponds to the *<start>* node. It has two children, $c_1^{(x)}$ and $c_2^{(x)}$, which correspond to the nodes *<preprocessing>* and *<classification>*, respectively. We define a function $ch_1(N)$ that returns the first child of node N. In this example, the *<imputation>* group is denoted as $ch_1(c_1^{(x)})$, whereas *<bernoulliNB>* is given by $ch_1(c_2^{(x)})$.

The distance between two trees T_x and T_y will initially depend on whether they include preprocessing steps or only a classification algorithm. Equation 4 shows four possible cases: neither T_x nor T_y have preprocessing steps, and only the distance from the classification algorithm ($dist_{clf}$) is accounted for (**C1**); both trees have preprocessing steps, and we calculate the distances from the two sides of the tree ($dist_{pre}$ and $dist_{clf}$) (**C2**); only T_x (**C3**) or T_y (**C4**) have a preprocessing step, so we calculate the distance between the classification subtrees and add a constant k to the distance, where $k = d(\text{<preprocessing>}, NULL)$ is the distance between the *<preprocessing>* non-terminal and the $NULL$ symbol, which is greater than the distance between any two preprocessing algorithms.

$$dist(T_x, T_y) = \begin{cases} dist_{clf}(ch_1(c_1^{(x)}), ch_1(c_1^{(y)})) & \text{C1} \\ dist_{pre}(c_1^{(x)}, c_1^{(y)}) + dist_{clf}(ch_1(c_2^{(x)}), ch_1(c_2^{(y)})) & \text{C2} \\ k + dist_{clf}(ch_1(c_1^{(x)}), ch_1(c_2^{(y)})) & \text{C3} \\ k + dist_{clf}(ch_1(c_2^{(x)}), ch_1(c_1^{(y)})) & \text{C4} \end{cases} \quad (4)$$

To the best of our knowledge, the way we calculate the distance between two preprocessing subtrees cannot be expressed in closed form, and function $dist_{pre}$ is described in Algorithm 1, where $children(N)$ is a function that returns all the children of node N. Sets A and B are initialized with the preprocessing groups of trees T_x and T_y, respectively (line 3). The first component of the distance is calculated as the distance from all groups that are only present in one of the trees to the $NULL$ symbol (lines 4 and 5). The second component consists of the distances between groups that are present in both trees (line 6). The loop in lines 7–16 compares the groups in the intersection. Function $get\_node(l, X)$ returns the node in set X whose label is l. For each group, if the algorithms in trees T_x and T_y are different, we add their distance to the total distance (line 11). If they are the same, we add to the total distance the distance between the values of their hyperparameters (lines 13 and 14). The total distance is then returned in line 17.

As an example, consider that Algorithm 1 receives the pipelines from Figs. 1 and 2b as T_x and T_y, respectively. In line 3, set A receives node *<imputation>*

and B receives nodes <imputation> and <dimensionality>. Thus, the difference between the two sets is composed of node <dimensionality> and its distance to the *NULL* symbol is added to the total distance. In line 4, *intersect* gets the symbol <imputation> and the corresponding nodes in trees T_x and T_y are retrieved in lines 8 and 9. $algA$ and $algB$ correspond to the same algorithm, so the distances between the values of their hyperparameters are added to the total distance.

Algorithm 1. Distance between preprocessing subtrees

1: **procedure** DIST$_{pre}$(T_x, T_y)   ▷ The root of T_x and T_y is the <preprocessing> symbol
2:     $distance \leftarrow 0$
3:     $A \leftarrow \{n \mid n \in children(T_x)\}$; $B \leftarrow \{n \mid n \in children(T_y)\}$
4:     $diff \leftarrow (A - B) \cup (B - A)$
5:     $distance \leftarrow distance + \sum_{i=1}^{|diff|} d(diff_i, NULL)$
6:     $intersect \leftarrow A \cap B$
7:     **for all** $label \in intersect$ **do**
8:         $algA \leftarrow ch_1(get\_node(label, A))$
9:         $algB \leftarrow ch_1(get\_node(label, B))$
10:        **if** $algA \neq algB$ **then**   ▷ Different algorithms
11:            $distance \leftarrow distance + d(algA, algB)$
12:        **else**   ▷ Same algorithm; check hyperparameters
13:            $hpA \leftarrow children(algA)$; $hpB \leftarrow children(algB)$
14:            $distance \leftarrow distance + \sum_{i=2}^{|hpA|} d(ch_1(hpA_i), ch_1(hpB_i))$
15:        **end if**
16:    **end for**
17:    **return** $distance$
18: **end procedure**

Equation 5 handles the case of the distance between the classification algorithms. In the first case of the equation, the algorithms are the same. Thus, the roots r_1 and r_2, which correspond to the non-terminals with the names of the algorithms, have the same number of children, m. In this case, the distance between the trees is the summation of the distance between the values of the hyperparameters of each algorithm. If the algorithms are the same, the distance between the trees is the distance between the algorithms, given in Table 1.

$$dist_{clf}(T_x, T_y) = \begin{cases} \sum_{j=2}^{m} d(ch_1(c_j^{(x)}), ch_1(c_j^{(y)})) & \text{if } r_x = r_y, \\ d(r_x, r_y) & \text{otherwise.} \end{cases} \qquad (5)$$

3.3   Fitness Function

The final component of the fitness landscape is a fitness function $f : X \rightarrow \mathbb{R}$ that maps each element of the set X of configurations to a real number. Here we deal with multiclass classification problems with class imbalance. Therefore, we defined the fitness function as the weighted F-measure [23] to evaluate the pipeline's learning model on the dataset of interest.

F-measure is defined in Eq. 6 for binary classification problems, where TP, FP and FN are the number of true positives, false positives and false negatives of the pipeline's learning model on the dataset, respectively.

$$F\text{-measure} = \frac{2 \cdot TP}{2 \cdot TP + FP + FN}, \tag{6}$$

As we deal with multiclass classification problems, we use a one-vs-all approach, transforming a problem of c classes into c binary classification problems to calculate the F-measure. We then calculate a weighted average of the F-measure over the c binary classification problems.

4   AutoML Fitness Landscape Analysis

For continuous optimization problems with two variables, fitness landscapes can be visualized, where the XY plane is the search space and fitness represents the third dimension. However, given the complexity of real world problems, this approach is not feasible. Several metrics have been proposed to describe and compare fitness landscapes for different problems. These metrics evaluate a number of features of optimization problems that play a role in the performance of search algorithms, such as modality, fitness distribution in the search space, ruggedness, degree of variable interdependency, evolvability, neutrality, among others [14, 19].

After we have defined the fitness landscape of AutoML problems, we will use these metrics to perform an analysis of the characteristics of this space. We focus on two metrics: the fitness distance correlation (FDC) and the mean neutrality ratio of the landscape. FDC is a popular way of measuring how the fitness function correlates with the distance to the global optimum, which is a way of measuring problem difficulty [9]. Neutrality, on the other hand, indicates the presence of regions in the search space with equal (or nearly equal, in the case of continuous spaces) fitness function values, which can have positive or negative impacts on the performance of optimization algorithms [14].

Fitness Distance Correlation: The fitness distance correlation (FDC) measure was proposed by the authors in [9] to give a global view of problem difficulty for genetic algorithms, but it has been frequently used as a metric to evaluate the fitness landscape of other optimization problems [14]. In its original formulation, FDC requires knowledge of the global optimum, which is unfeasible for AutoML problems. The authors in [10] proposed an adaption of FDC, called FDC_s, for continuous problems with no known global optimum. Given a sample of n points $X = \{x_1, \ldots, x_n\}$ from the search space with associated fitness values $F = \{f_1, \ldots, f_n\}$ with mean $\overline{f}$, the best point in the sample is denoted by x^*. The Euclidean distance from x^* to every point $x_i \in X$ is denoted by $D^* = \{d_1^*, \ldots, d_n^*\}$, with mean $\overline{d^*}$. FDC_s is given by Eq. 7. From here on, we denote FDC_s simply by FDC.

$$FDC_s = \frac{\sum_{i=1}^{n}(f_i - \overline{f})(d_i^* - \overline{d^*})}{\sqrt{\sum_{i=1}^{n}(f_i - \overline{f})^2}\sqrt{\sum_{i=1}^{n}(d_i^* - \overline{d^*})^2}} \tag{7}$$

FDC returns a value between -1 (perfect anti-correlation) and +1 (perfect correlation). For maximization problems, search spaces with low FDC values

Table 2. Datasets used in the experiments.

Dataset	Instances	Features	Classes	Missing	Source
breast-w	699	9	2	Yes	OpenML
diabetes	768	8	2	No	OpenML
stalog-segment	2310	19	7	No	UCI
vehicle	846	18	4	No	OpenML
wilt	4839	5	2	No	OpenML
wine-quality-red	1599	11	6	No	OpenML

are considered easy, values around 0 are difficult, and high values correspond to misleading spaces [9,10]. Given the nature of the AutoML search space, we replace the Euclidean distances D^* used in Eq. 7 by the distance measure defined in Eq. 4.

Neutrality: Neutrality identifies the presence of regions in the landscape with equal or similar fitness [17]. Its role in determining the ability of an optimization method to find good solutions has been a topic of discussion, especially in the context of evolutionary algorithms. There is evidence that neutrality can both make the search space easier to explore [21] or get some algorithms stuck in regions of the search space with equal fitness, preventing them from exploring areas with possibly better results [14].

In the context of AutoML, we define a neutral neighborhood $N^\approx(s)$ of a solution s as the set $N^\approx(s) = \{s' \in N(s) \mid |f(s') - f(s)| < \delta\}$ for some small constant $\delta \geq 0$, where $f(s)$ is the fitness of s and $N(s)$ is a sample of the complete neighborhood of s, as defined in Sect. 3.2.

The cardinality of $N^\approx(s)$ is called the neutrality degree of s, whereas the neutrality ratio of s is given by $|N^\approx(s)|/|N(s)|$ [21]. These metrics give us an overview of the neutrality level of the landscape.

5   Experimental Analysis

In this section, we present the FDC and neutrality results for six classification datasets, obtained either from UCI [2] or OpenML [22]. Table 2 summarizes their main characteristics, including the number of instances, features and classes, the presence or absence of missing values, and the data source. Correlation analyses are reported considering Spearman's rank correlation coefficient (ρ) and the two-sided p-value for the hypothesis test (null hypothesis is that the two sets of data are uncorrelated) [25].

All pipeline configurations were generated using algorithms implemented in the Python library Scikit-learn [13] and evaluated using 5-fold cross-validation. All results reported correspond to an average of 30 independent samples from the search space.

5.1   Fitness Distance Correlation Analysis

For the FDC analysis, we generated 30 random samples of the search space of varying sizes, ranging from 500 to 3,000 in intervals of 500. The pipelines in each sample were generated by randomly selecting production rules from the grammar.

Figure 3 shows the FDC values for different sample sizes. Observe that increasing the sample size has little effect on FDC, showing our sample is able to capture the overall trend of this metric in the evaluated search space. We found a slight positive correlation between FDC and the mean fitness ($\rho = 0.222$, $p < 0.01$), which is somewhat unexpected, since higher FDC values (in our case, values that are closer to 0) should be related to harder problems.

We tried to find a relation between the mean values of accuracy reported in the OpenML repository for the datasets used in the experiments and the values of FDC. For example, *breast-w* has mean accuracy of 0.93 and FDC values always smaller than -0.2. *Diabetes*, in turn, has a mean accuracy of 0.75 and values of FDC always in the interval $[-0.05, 0]$. However, this is not consistent for all datasets. *Wilt*, for instance, has an accuracy of 0.98 and FDC values as small as *diabetes*. Analysis of FDC for a greater number of datasets and looking at their main characteristics are yet to be performed. Further, FDC results are correlated with the distance measure used in this work, and a more in-depth evaluation of this measure is also subject of future work.

However, for AutoML problems, there may be other factors related to low fitness apart from the difficulty of the problem. In order to fully understand the relation between FDC and problem difficulty in the context of AutoML, further experimentation varying the search space is necessary.

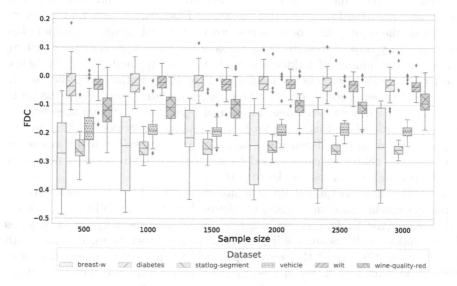

Fig. 3. FDC values for different sample sizes.

5.2 Neutrality Analysis

For the analysis of the neutrality of the search space, we performed random walks starting from a random position. For each point of the walk (solution), we evaluated the fitness function of a given number of neighbors and analyzed the neutrality ratio. One of the neighbors was then selected as the next starting point of the walk. Here we report the mean neutrality ratio of the complete walk.

Recall that, for continuous spaces, we consider regions that are neutral within a specified tolerance δ for the fitness difference between two solutions. For each dataset, we defined δ as the standard deviation of the mean fitness of 30 independent random samples of size 1000.

Figure 4 shows the neutrality ratios for different walk lengths (100, 200, 300 and 400) and neighborhood sizes (5, 10, 15 and 20). The tolerance δ for each dataset is shown in the title of the corresponding plot. As we can see, increasing the walk length and the neighborhood size does not have a strong impact on the average neutrality ratio, but the variance decreases. We found a strong positive correlation between the mean neutrality ratio and the fitness value ($\rho = 0.715$, $p = 0.0$). This result shows higher neutrality in regions of the search space with higher fitness values. However, in order to understand how neutrality affects the performance of different AutoML optimization methods, further experimentation with variations of the search space and its neutrality ratio are also necessary.

6 Related Work

Fitness landscape analysis has been vastly explored for typical optimization problems [14], but the literature regarding such analysis for machine learning problems is scarce. Much of the effort has been directed to neural network error landscapes. Rakitianskaia et al. [16] measured FDC, ruggedness and gradients to evaluate the error landscape of fully-connected neural networks used for classification. They showed that the ruggedness of the landscapes decreases with an increase in the number of hidden layers of the network, making the landscape "harder" to explore, whereas the results for FDC indicate that it can be used to determine the searchability of different network architectures for specific problems. Another study analyzed neutrality in such landscapes, whose presence can hinder population-based methods for training neural networks [1]. The authors proposed two measures of neutrality based on random walks on the landscape and suggested that they can be used to study the relation between neutrality and the performance of search algorithms.

In [4], the authors explored different subsets of the unbounded neural network search space using random walks. They found high-magnitude fitness gradients and more rugged landscapes for larger search spaces, specially for large steps of the random walk. Searchability metrics, on the other hand, decrease with an increase in the size of the search space. These properties reflect a greater difficulty in searching larger spaces. A subsequent study by the same group proposed a progressive random walk method for sampling network error landscapes [3]. The authors noted that methods based on random walks may not

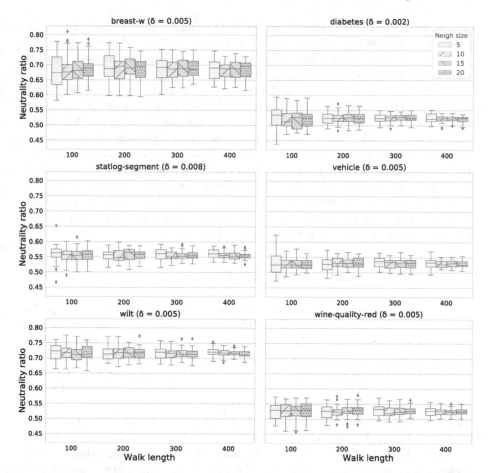

Fig. 4. Neutrality ratios for varying walk lengths and neighborhood sizes. The tolerance δ for each dataset is shown in parentheses.

cover regions with high fitness values, thus not representing the search space well. The results showed that the proposed method is more computationally efficient than population-based walks and is very successful in finding areas of high fitness.

The landscape of the algorithm configuration problem has been evaluated in [15] in terms of the modality and convexity of parameter responses. The authors defined parameter response slices by parameter p within a given window around an optima found by the Sequential Model-based Algorithm Configuration (SMAC), keeping all other parameters fixed and measuring the performance of the algorithm as a function of p. This procedure is repeated for all parameters being considered. They evaluated algorithms for three typical optimization problems, namely SAT, MIP and TSP, and concluded that many of the parameter slices appear to be uni-modal and convex, both on instance sets and on individual instances, although the former leads to a more rugged landscape. This algorithm configuration analysis is related to the problem of hyperparameter optimization of machine learning algorithms, but it does not consider neither algorithm selection nor categorical parameters.

Garcianera *et al.* [7], in turn, performed an analysis of a subset of the search space explored by TPOT [12], an AutoML tool that uses genetic programming to evolve machine learning pipelines for regression and classification problems. The authors defined a neighborhood relation in which two pipelines are neighbors if they differ in a single algorithm or parameter. Using a reduced grid search, they compared the classification accuracy of TPOT with stochastic, random-restart hill climbing and random search. The results suggest the existence of several regions with high fitness, but which are prone to overfitting. However, the paper fails to analyze other characteristics of the fitness landscape and how they can influence the performance of optimization methods.

7   Conclusions and Future Work

The main contribution of this paper is the definition of a fitness landscape for AutoML problems. We proposed a flexible representation for machine learning pipelines that captures the relative importance of changing an algorithm by another or modifying the value of a hyperparameter. We use this representation to define a notion of neighborhood and the distance between pipelines. We found a strong correlation between the mean fitness ratio and fitness values, and a high correlation between fitness values and neutrality.

Having defined the components of the AutoML search space, the next steps include modifying the search space to evaluate how the metrics change in response to the size of the space. We also plan on testing other sampling strategy to take into account the differences in the size of the search space induced by categorical, discrete and continuous hyperparameters. Another possible direction of future work is analysing how different AutoML optimization methods behave in the presence of different levels of neutrality and for different FDC values.

Acknowledgments. The UFMG authors would like to thank FAPEMIG, CNPq and CAPES for their financial support. This work has also been partially funded by ATMO-SPHERE (H2020 777154 and MCTIC/RNP 51119).

References

1. van Aardt, W.A., Bosman, A.S., Malan, K.M.: Characterising neutrality in neural network error landscapes. In: Proceedings of the Congress on Evolutionary Computation, pp. 1374–1381. IEEE (2017)
2. Asuncion, A., Newman, D.: UCI machine learning repository (2007)
3. Bosman, A.S., Engelbrecht, A.P., Helbig, M.: Progressive gradient walk for neural network fitness landscape analysis. In: Proceedings of the Genetic and Evolutionary Computation Conference, pp. 1473–1480. ACM (2018)

4. Bosman, A.S., Engelbrecht, A., Helbig, M.: Search space boundaries in neural network error landscape analysis. In: Proceedings of the Symposium Series on Computational Intelligence, pp. 1–8. IEEE (2016)
5. Ekárt, A., Németh, S.Z.: A metric for genetic programs and fitness sharing. In: Poli, R., Banzhaf, W., Langdon, W.B., Miller, J., Nordin, P., Fogarty, T.C. (eds.) EuroGP 2000. LNCS, vol. 1802, pp. 259–270. Springer, Heidelberg (2000). https://doi.org/10.1007/978-3-540-46239-2_19
6. Feurer, M., Klein, A., Eggensperger, K., Springenberg, J., Blum, M., Hutter, F.: Efficient and robust automated machine learning. In: Proceedings of the Advances in Neural Information Processing Systems, pp. 2962–2970 (2015)
7. Garciarena, U., Santana, R., Mendiburu, A.: Analysis of the complexity of the automatic pipeline generation problem. In: Proceedings of the Congress on Evolutionary Computation, pp. 1–8. IEEE (2018)
8. Hutter, F., Kotthoff, L., Vanschoren, J. (eds.): Automated Machine Learning. TSSCML. Springer, Cham (2019). https://doi.org/10.1007/978-3-030-05318-5. http://automl.org/book
9. Jones, T., Forrest, S.: Fitness distance correlation as a measure of problem difficulty for genetic algorithms. In: Proceedings of the 6th International Conference on Genetic Algorithms, pp. 184–192 (1995)
10. Malan, K.M., Engelbrecht, A.P.: Characterising the searchability of continuous optimisation problems for PSO. Swarm Intell. **8**(4), 275–302 (2014). https://doi.org/10.1007/s11721-014-0099-x
11. Mckay, R.I., Hoai, N.X., Whigham, P.A., Shan, Y., O'Neill, M.: Grammar-based genetic programming: a survey. Genet. Program Evolvable Mach. **11**(3–4), 365–396 (2010). https://doi.org/10.1007/s10710-010-9109-y
12. Olson, R.S., Moore, J.H.: TPOT: a tree-based pipeline optimization tool for automating machine learning. In: Hutter, F., Kotthoff, L., Vanschoren, J. (eds.) Automated Machine Learning. TSSCML, pp. 151–160. Springer, Cham (2019). https://doi.org/10.1007/978-3-030-05318-5_8
13. Pedregosa, F., et al.: Scikit-learn: machine learning in python. JMLR **12**(Oct), 2825–2830 (2011)
14. Pitzer, E., Affenzeller, M.: A comprehensive survey on fitness landscape analysis. In: Klempous, R., Suárez Araujo, C.P. (eds.) Recent Advances in Intelligent Engineering Systems, vol. 378, pp. 161–191. Springer, Heidelberg (2012). https://doi.org/10.1007/978-3-642-23229-9_8
15. Pushak, Y., Hoos, H.: Algorithm configuration landscapes: In: Auger, A., Fonseca, C.M., Lourenço, N., Machado, P., Paquete, L., Whitley, D. (eds.) PPSN 2018. LNCS, vol. 11102, pp. 271–283. Springer, Cham (2018). https://doi.org/10.1007/978-3-319-99259-4_22
16. Rakitianskaia, A., Bekker, E., Malan, K.M., Engelbrecht, A.: Analysis of error landscapes in multi-layered neural networks for classification. In: Proceedings of the 2016 IEEE Congress on Evolutionary Computation, pp. 5270–5277. IEEE (2016)
17. Reidys, C.M., Stadler, P.F.: Neutrality in fitness landscapes. Appl. Math. Comput. **117**(2–3), 321–350 (2001). https://doi.org/10.1016/S0096-3003(99)00166-6
18. Sipser, M.: Introduction to the Theory of Computation. 3rd edn. Cengage Learning (2012)
19. Stadler, P.F.: Fitness landscapes. In: Lässig, M., Valleriani, A. (eds.) Biological Evolution and Statistical Physics, vol. 585, pp. 183–204. Springer, Heidelberg (2002). https://doi.org/10.1007/3-540-45692-9_10

20. Thornton, C., Hutter, F., Hoos, H.H., Leyton-Brown, K.: Auto-WEKA: combined selection and hyperparameter optimization of classification algorithms. In: Proceedings of the 19th ACM SIGKDD International Conference on Knowledge Discovery and Data Mining, pp. 847–855. ACM (2013)
21. Vanneschi, L., Pirola, Y., Mauri, G., Tomassini, M., Collard, P., Verel, S.: A study of the neutrality of boolean function landscapes in genetic programming. Theor. Comput. Sci. **425**, 34–57 (2012)
22. Vanschoren, J., Van Rijn, J.N., Bischl, B., Torgo, L.: OpenML: networked science in machine learning. ACM SIGKDD Explor. Newsl. **15**(2), 49–60 (2014)
23. Witten, I.H., Frank, E., Hall, M.A., Pal, C.J.: Data mining: practical machine learning tools and techniques, 4th edn. Morgan Kaufmann Publishers Inc., Burlington (2016)
24. Zöller, M.A., Huber, M.F.: Survey on automated machine learning. arXiv preprint arXiv:1904.12054 (2019)
25. Zwillinger, D.: CRC standard mathematical tables and formulae. Chapman and Hall/CRC, London/Boca Raton (2002)

On the Combined Impact of Population Size and Sub-problem Selection in MOEA/D

Geoffrey Pruvost[1(✉)], Bilel Derbel[1(✉)], Arnaud Liefooghe[1(✉)], Ke Li[2(✉)], and Qingfu Zhang[3(✉)]

[1] University of Lille, CRIStAL, Inria, Lille, France
{geoffrey.pruvost,bilel.derbel,arnaud.liefooghe}@univ-lille.fr
[2] University of Exeter, Exeter, UK
k.li@exeter.ac.uk
[3] City University Hong Kong, Kowloon Tong, Hong Kong
qingfu.zhang@cityu.edu.hk

Abstract. This paper intends to understand and to improve the working principle of decomposition-based multi-objective evolutionary algorithms. We review the design of the well-established MOEA/D framework to support the smooth integration of different strategies for sub-problem selection, while emphasizing the role of the population size and of the number of offspring created at each generation. By conducting a comprehensive empirical analysis on a wide range of multi- and many-objective combinatorial NK landscapes, we provide new insights into the combined effect of those parameters on the anytime performance of the underlying search process. In particular, we show that even a simple random strategy selecting sub-problems at random outperforms existing sophisticated strategies. We also study the sensitivity of such strategies with respect to the ruggedness and the objective space dimension of the target problem.

1 Introduction

Context. Evolutionary multi-objective optimization (EMO) algorithms [7] have been proved extremely effective in computing a high-quality approximation of the Pareto set, i.e., the set of solutions providing the best trade-offs among the objectives of a multi-objective combinatorial optimization problem (MCOP). Since the working principle of an evolutionary algorithm (EA) is to evolve a *population* of solutions, this population can be explicitly mapped with the target approximation set. The goal is then to improve the quality of the population, and to guide its incumbent individuals to be as close and as diverse as possible w.r.t. the (unknown) Pareto set. Existing EMO algorithms can be distinguished according to how the population is evolved. They are based on an iterative process where at each iteration: (i) some individuals (parents) from the population are selected, (ii) new individuals (offspring) are generated using variation operators (e.g., mutation, crossover) applied to the selected parents,

© Springer Nature Switzerland AG 2020
L. Paquete and C. Zarges (Eds.): EvoCOP 2020, LNCS 12102, pp. 131–147, 2020.
https://doi.org/10.1007/978-3-030-43680-3_9

and (iii) a replacement process updates the population with newly generated individuals. Apart from the problem-dependent variation operators, the design of selection and replacement is well-understood to be the main challenge for an efficient and effective EMO algorithm, since these interdependent steps allow to control both the convergence of the population and its diversity. In contrast with dominance- (e.g., [7]) or indicator-based (e.g., [2]) approaches, aggregation-based approaches [16] rely on the transformation of the objective values of a solution into a scalar value, that can be used for selection and replacement. In this paper, we are interested in studying the working principles of this class of algorithms, while focusing on the so-called MOEA/D (Multi-objective evolutionary algorithm based on decomposition) [11,22], which can be considered as a state-of-the-art framework.

Motivations. The MOEA/D framework is based on the decomposition of the original MCOP into a set of smaller sub-problems that are mapped to a population of individuals. In its basic variant [22], MOEA/D considers a set of single-objective sub-problems defined using a scalarizing function transforming a multi-dimensional objective vector into a scalar value w.r.t. one weight (or direction) vector in the objective space. The population is then typically structured by mapping one individual to one sub-problem targeting a different region of the objective space. Individuals from the population are evolved following a cooperative mechanism in order for each individual (i) to optimize its own sub-problem, and also (ii) to help solving its neighboring sub-problems. The population hence ends up having a good quality w.r.t. all sub-problems. Although being extremely simple and flexible, the computational flow of MOEA/D is constantly redesigned to deal with different issues. Different MOEA/D variants have been proposed so far in the literature, e.g., to study the impact of elitist replacements [19], of generational design [13], or of stable-matching based evolution [12], and other mechanisms [1]. In this paper, we are interested in the interdependence between the population size, which is implied by the number of sub-problems defined in the initial decomposition, and the internal evolution mechanisms of MOEA/D.

The population size has a deep impact on the dynamics and performance of EAs. In MOEA/D, the sub-problems target diversified and representative regions of the Pareto front. They are usually defined to spread evenly in the objective space. Depending on the shape of the (unknown) Pareto front, and on the number of objectives, one may need to define a different number of sub-problems. Since, the population is structured following the so-defined sub-problems, it is not clear how the robustness of the MOEA/D selection and replacement strategies can be impacted by a particular setting of the population size. Conversely, it is not clear what population size shall be chosen, and how to design a selection and replacement mechanism implying a high-quality approximation. Besides, the proper setting of the population size (see e.g. [6,8,20]) in is EAs can depend on the problem properties, for example in terms of solving difficulty. EMO algorithms are no exceptions. In MOEA/D, sub-problems may have different characteristics, and the selection and replacement mechanisms can be guided by such considerations. This is for example the case for a number of MOEA/D variants where it is argued that some sub-problems might be more difficult to solve than others [18,23], and hence that the population shall be guided accordingly.

Methodology and Contribution. In this paper, we rely on the observation that the guiding principle of MOEA/D can be leveraged in order to support a simple and high-level tunable design of the selection and replacement mechanisms on one hand, while enabling a more fine-grained control over the choice of the population size, and subsequently its impact on approximation quality on the other hand. More specifically, our work can be summarized as follows:

- We consider a revised design of MOEA/D which explicitly dissociates between three components: (i) the number of individuals selected at each generation, (ii) the strategy adopted for selecting those individuals and (iii) the setting of the population size. Although some sophisticated strategies to distribute the computational effort of sub-problems exploration were integrated within some MOEA/D variants [10,18,23], to the best of our knowledge, the individual impact of such components were loosely studied in the past.
- Based on this fine-grained revised design, we conduct a comprehensive analysis about the impact of those three components on the convergence profile of MOEA/D. Our analysis is conducted in an incremental manner, with the aim of providing insights about the interdependence between those design components. In particular, we show evidence that the number of sub-problems selected at each generation plays an even more important role than the way the sub-problems are selected. Sophisticated selection strategies from the literature are shown to be outperformed by simpler, well configured strategies.
- We consider a broad range of multi- and many-objective NK landscapes, viewed as a standard and difficult family of MCOP benchmarks, which is both scalable in the number of objectives and exposes a controllable difficulty in terms of ruggedness. By a thorough benchmarking effort, we are then able to better elicit the impact of the MOEA/D population size, and the robustness of selection strategies on the (anytime) approximation quality.

It is worth noticing that our work shall not be considered as yet another variant in the MOEA/D literature. In fact, our analysis precisely aims at enlightening the main critical design parameters and components that can be hidden behind a successful MOEA/D setting. Our investigations are hence to be considered as a step towards the establishment of a more advanced component-wise configuration methodology allowing the setting up of future high-quality decomposition-based EMO algorithms for both multi- and many-objective optimization.

Outline. In Sect. 2, we recall basic definitions and we detail the working principle of MOEA/D. In Sect. 3, we describe our contribution in rethinking MOEA/D by explicitly dissociating between the population size and the number of selected sub-problems, then allowing us to leverage existing algorithms as instances of the revised framework. In Sect. 4, we present our experimental study and we state our main findings. In Sect. 5, we conclude the paper and discuss further research.

2   Background

2.1   Multi-objective Combinatorial Optimization

A *multi-objective combinatorial optimization problem* (MCOP) can be defined by a set of M objective functions $f = (f_1, f_2, \ldots, f_M)$, and a discrete set X of feasible solutions in the *decision space*. Let $Z = f(X) \subseteq \mathbb{R}^M$ be the set of feasible outcome vectors in the *objective space*. To each solution $x \in X$ is assigned an objective vector $z \in Z$, on the basis of the vector function $f : X \to Z$. In a maximization context, an objective vector $z \in Z$ is *dominated* by a vector $z' \in Z$ iff $\forall m \in \{1, \ldots, M\}$, $z_m \leqslant z'_m$ and $\exists m \in \{1, \ldots, M\}$ s.t. $z_m < z'_m$. A solution $x \in X$ is dominated by a solution $x' \in X$ iff $f(x)$ is dominated by $f(x')$. A solution $x^\star \in X$ is *Pareto optimal* if there does not exist any other solution $x \in X$ such that $x^\star$ is dominated by x. The set of all Pareto optimal solutions is the *Pareto set*. Its mapping in the objective space is the *Pareto front*. The size of the Pareto set is typically exponential in the problem size. Our goal is to identify a good *Pareto set approximation*, for which EMO algorithms constitute a popular effective option [7]. As mentioned before, we are interested in aggregation-based methods, and especially in the MOEA/D framework which is sketched below.

2.2   The Conventional MOEA/D Framework

Aggregation-based EMO algorithms seek good-performing solutions in multiple regions of the Pareto front by *decomposing* the original multi-objective problem into a number of *scalarized* single-objective *sub-problems* [16]. In this paper, we use the Chebyshev scalarizing function: $\mathbf{g}(x, \omega) = \max_{i \in \{1, \ldots, M\}} \omega_i \cdot \left| z_i^\star - f_i(x) \right|$, where $x \in X$, $\omega = (\omega_1, \ldots, \omega_M)$ is a positive weight vector, and $z^\star = (z_1^\star, \ldots, z_M^\star)$ is a reference point such that $z_i^\star > f_i(x) \; \forall x \in X$, $i \in \{1, \ldots, M\}$.

In MOEA/D [22], sub-problems are optimized cooperatively by defining a *neighborhood relation* between sub-problems. Given a set of μ weight vectors $\mathcal{W}_\mu = (\omega^1, \ldots, \omega^\mu)$, with $\omega^j = (\omega_1^j, \ldots, \omega_M^j)$ for every $j \in \{1, \ldots, \mu\}$, defining μ sub-problems, MOEA/D maintains a population $P_\mu = (x^1, \ldots, x^\mu)$ where each individual x^j corresponds to one sub-problem. For each sub-problem $j \in \{1, \ldots, \mu\}$, a set of neighbors $\mathcal{B}_j$ is defined by considering the T closest weight vectors based on euclidean distance. All sub-problems are considered at each generation. Given a sub-problem j, two sub-problems are selected at random from $\mathcal{B}_j$, and the two corresponding solutions are considered as parents. An offspring x' is created by means of variation (e.g., crossover, mutation). For every $k \in \mathcal{B}_j$, if x' improves k's current solution x^k, then x' replaces it, i.e., if $\mathbf{g}(x', \omega^k) < \mathbf{g}(x^j, \omega^k)$ then $x^k = x'$. The algorithm loops over sub-problems, i.e., weight vectors, or equivalently over the individuals in the population, until a stopping condition is satisfied. In the conventional MOEA/D terminology, an iteration refers to making selection, offspring generation, and replacement for *one* sub-problem. By contrast, a generation consists in processing all sub-problems once, i.e., after one generation μ offspring are generated. Notice that other issues are also addressed, such as the update of the reference point z^* required by the scalarizing function, and the option to incorporate an external archive for storing all non-dominated points found so far during the search process.

From the previous description, it should be clear that, at each iteration, MOEA/D is applying an elitist $(T+1)$-EA w.r.t. the sub-population $\mathcal{B}_i$ underlying the neighborhood of the current sub-problem. After one generation, one can roughly view MOEA/D as applying a $(\mu + \mu)$-EA w.r.t. the full population. A noticeable difference is that the basic MOEA/D is not a generational algorithm, in the sense that it does not handle the population as a whole, but rather in a local and greedy manner. This is actually a distinguishable feature of MOEA/D, since the population is structured by the initial sub-problems and evolved accordingly.

3 Revising and Leveraging the Design of MOEA/D

3.1 Positioning and Rationale

As in any EA, both the population size and the selection and replacement mechanisms of MOEA/D play a crucially important role. Firstly, a number of weight, i.e., a population size, that is too small may not only be insufficient to cover well the whole Pareto front, but may also prevent the identification of high-quality solutions for the defined sub-problems. This is because the generation of new off-spring is guided by the so-implied $(T+1)$-EA for which the local sub-population of size T might be too different and hence too restrictive for generating good offspring. On the other hand, a too large population may result in a substantial waste of resources, since too many sub-problems might map to the same solution. Secondly, a small population size can be sufficient to approach the Pareto front in a reduced number of steps. However, a larger population is preferable to better cover the Pareto front. As in single-objective optimization, a larger population might also help escaping local optima [20]. As a result, it is not clear what is a proper setting of the population size in MOEA/D, since the previously discussed issues seem contradictory.

Although one can find different studies dealing with the impact of the population size in EAs [5,6,8,20], this issue is explicitly studied only to a small extent, especially for decomposition-based multi- and many-objective optimization [9,15]. For instance, in [8], offline and online scheduling strategies for controlling the population size are coupled with SMS-EMOA [2], a well-known indicator-based EMO algorithm, for bi-objective continuous benchmarks. Leveraging such a study to combinatorial domains with more than two objectives, and within the MOEA/D framework, is however a difficult question. Tightly related to the population size, other studies investigate the distribution of the computational effort over the sub-problems [3,4,10,18,23]. The rationale is that the defined sub-problems might have different degrees of difficulty and/or that the progress over some sub-problems might be more advanced than others in the course of the search process. Hence, different adaptive mechanisms have been designed in order to detect which sub-problems to consider, or equivalently which solutions to select when generating a new offspring. A representative example of such approaches is the so-called MOEA/D–DRA (MOEA/D with dynamical resource allocation) [23], that can be considered as a state-of-the-art algorithm when dealing with the proper distribution of the computational effort over sub-problems. In MOEA/D–DRA, a utility function is defined w.r.t. the current status of sub-problems. A tournament selection is used to decide which sub-problems to select when generating a new offspring. Despite a skillful design, such an approach stays focused on the relative difficulty of solving sub-problems, while omitting to analyze the impact of the number of selected sub-problems and its interaction with both the population size and the characteristics of the underlying problem.

We propose to revise the MOEA/D framework and to study in a more explicit and systematic manner the combined effect of population size and sub-problem selection in light of the properties of the MCOP at hand. As mentioned above, this was investigated only to a small extent in the past although, as revealed by our experimental findings, it is of critical importance to reach an optimal performance when adopting the MOEA/D framework.

3.2   The Proposed MOEA/D–(μ, λ, sps) Framework

In order to better study and analyze the combined effect of population size and sub-problem selection, we propose to rely on a revised framework for MOEA/D, denoted MOEA/D–(μ, λ, sps), as defined in the high-level template depicted in Algorithm 1. This notation is inspired by the standard ($\mu + \lambda$)-EA scheme, where starting from a population of size μ, λ new individuals are generated and merged to form a new population of size μ after replacement. In the MOEA/D framework, however, this has a specific meaning as detailed in the following.

The proposed MOEA/D–(μ, λ, sps) algorithm follows the same steps as the original MOEA/D. However, it explicitly incorporates an additional component, denoted sps, which stands for the sub-problem selection strategy. Initially, the population is generated and mapped to the initial μ weight vectors. An optional external archive is also incorporated in the usual way with no effect on the search process. The algorithm then proceeds in different generations (the outer while loop). At each generation, λ sub-problems, denoted I_λ, are selected using the sps strategy. A broad range of deterministic and stochastic selection strategies can be integrated. In particular, λ can be though as an intrinsic parameter of the EMO algorithm itself, or implied by a specific sps strategy. The so-selected sub-problems are processed in order to update the population (the inner for loop). For the purpose of this paper, we adopt the same scheme than conventional MOEA/D: selected sub-problems are processed in an iterative manner, although other generational EA schemes could be adopted. At each iteration, that is for each selected sub-problem, denoted i, some parents are selected as usual from the T-neighborhood $\mathcal{B}_i$ of weight vector ω^i w.r.t. $\mathcal{W}_\mu$. The setting of the neighborhood $\mathcal{B}_i$ can be exactly the same as in conventional MOEA/D and

---

Algorithm 1. High level template of MOEA/D–(μ, λ, sps)

    Input: $\mathcal{W}_\mu := \{\omega^1, \ldots, \omega^\mu\}$: weights; $g(\cdot \mid \omega)$: scalar function; T: neighb. size;

1 $\mathcal{EP} \leftarrow \varnothing$: (optional) external archive ;

2 $\mathcal{P}_\mu \leftarrow \{x^1, \ldots, x^\mu\}$: generate and evaluate initial population of size μ;

3 $z^* \leftarrow$ initialize reference point from $\mathcal{P}_\mu$;

4 **while** *StoppingCriteria* **do**

5       $I_\lambda \leftarrow \mathsf{sps}(\mathcal{W}_\mu, \mathcal{P}_\mu, history)$;

6   **for** $i \in I_\lambda$ **do**

7           $\mathcal{B}_i \leftarrow$ the T-neighborhood of sub-problem i using $\mathcal{W}_\mu$;

8           $\mathcal{X} \leftarrow$ matingSelection($\mathcal{B}_i$);

9           $x' \leftarrow$ variation($\mathcal{X}$);

10         $F(x') \leftarrow$ evaluate x' ;

11         $\mathcal{EP} \leftarrow$ update external archive using x';

12         $z^* \leftarrow$ update reference point using $F(x')$;

13         $\mathcal{P}_\mu \leftarrow$ replacement($\mathcal{P}_\mu, x', \mathcal{B}_i \mid \mathbf{g}$);

14   *history* $\leftarrow$ update search history;

---

its variants. However, at this step, it is important to emphasize that the considered neighborhood $\mathcal{B}_i$ is w.r.t. the whole set of available weight vectors $\mathcal{W}_\mu$, that is considering all the initially designed sub-problems, and *not only* the selected ones. In particular, $\mathcal{B}_i$ may include some sub-problems that were *not* selected by the sps strategy. This is motivated by the fact that parents that are likely to produce a good offspring should be defined w.r.t. the population as a whole, and not solely within the subset of active sub-problems at a given generation, which might be restrictive. A new offspring x' is then generated using standard variation operators (e.g., crossover, mutation). The reference point required by the scalarizing function and the optional external archive are updated. Thereafter, the offspring is considered for replacement as in the conventional MOEA/D and its variants. Here again, this is handled using the neighborhood $\mathcal{B}_i$ of the current sub-problem i, computed w.r.t. the whole population. It is worth noticing that population update is made on the basis of the scalarizing function $\mathbf{g}$, which is a distinguishable feature of aggregation-based approaches.

At last, notice that we also use a *history* variable, referring to the evolution of the search state, and hence serving as a memory where any relevant information could be store for the future actions of the algorithm. In particular, we explicitly integrate the history within the sps strategy, since this will allow us to leverage some existing MOEA/D variants, as further discussed below.

3.3 Discussion and Outlook

It shall be clear from the previous description that the MOEA/D–(μ, λ, sps) framework allows us to emphasize the interdependence between three main components in a more fine-grained manner while following the same working principle than the original MOEA/D. Firstly, the number of weight vectors, or equivalently

Table 1. Different instantiations of the MOEA/D-$(\mu, \lambda, \mathsf{sps})$ framework.

Algorithm	Pop. size	# selected sub-prob.	Selection strategy	Ref.
MOEA/D	μ	μ	$\mathsf{sps_{ALL}}$	[22]
MOEA/D–DRA	μ	$\mu/5$	$\mathsf{sps_{DRA}}$	[23]
MOEA/D–RND	μ	$\lambda \leqslant \mu$	$\mathsf{sps_{RND}}$	here

the population size, is now made more explicit. In fact, the set of weight vectors now 'simply' plays the role of a global data structure to organize the individuals from the population. This structure can be used at the selection and replacement steps. In particular, one is not bound to iterate over all weight vectors, but might instead select a subset of individuals following a particular strategy. Secondly, the number of selected sub-problems λ determines directly the number of offspring to be generated at each generation. From an exploration/exploitation perspective, we believe this is of critical importance in general for $(\mu + \lambda)$-EAs, and it is now made more explicit within the MOEA/D framework. Furthermore, the λ offspring solutions are not simply generated from the individuals mapping to the selected sub-problems. Instead, parent selection interacts directly with the whole population, structured around the μ weight vectors, since the local neighborhood of each selected sub-problem may be used. Thirdly, the interaction between μ and λ is complemented more explicitly by the sub-problem selection strategy. In conventional MOEA/D for instance, the selection strategy turns out to be: $\mathsf{sps_{ALL}}$ = 'select all sub-problems', with $\lambda = \mu$. However, advanced MOEA/D variants can be captured as well. For instance, MOEA/D–DRA [23], focusing on the dynamic distribution of computations, can easily be instantiated as follows. For each sub-problem, we store and update the utility value as introduced in [23] by using the history variable. Let us recall that in MOEA/D–DRA, the utility of a sub-problem is simply the amount of progress made by solution x^i for sub-problem ω^i in terms of the scalarized fitness value $\mathsf{g}(\cdot|\omega^i)$ over different generations. In addition, M boundary weight vectors (in the objective space) are selected at each generation, and further $(\mu/5 - M)$ weight vectors are selected by means of a tournament selection of size 10. Hence, the sub-problem selection strategy turns out to be $\mathsf{sps_{DRA}}$ = 'select the boundary vectors and sub-problems using a tournament selection of size 10', with $\lambda = \mu/5$. Notice that this choice is to recall the one-fifth success rule from $(\mu + \lambda)$ evolution strategies [14].

In the reminder, MOEA/D–$(\mu, \mu, \mathsf{sps_{ALL}})$ refers to the conventional MOEA/D as described in [22], and MOEA/D–$(\mu, \mu/5, \mathsf{sps_{DRA}})$ refers to MOEA/D–DRA [23]; see Table 1. Other settings and parameters can be conveniently investigated as well. Since we are interested in the combined effect of μ, λ and sps, we also consider a simple baseline sub-problem selection strategy, denoted $\mathsf{sps_{RND}}$, which is to select a subset of sub-problems uniformly at random. Notice that our empirical analysis shall shed more lights on the behavior and the accuracy of the existing $\mathsf{sps_{DRA}}$ strategy.

4 Experimental Analysis

4.1 Experimental Setup

Multi-objective NK Landscapes. We consider multi-objective NK landscapes as a problem-independent model of multi-objective multi-modal combinatorial optimization problems [17]. Solutions are binary strings of size N and the objective vector to be maximized is defined as $f \colon \{0,1\}^N \mapsto [0,1]^M$. The parameter K defines the *ruggedness* of the problem, that is the number of (random) variables that influence the contribution of a given variable to the objectives. By increasing K from 0 to $(N-1)$, problems can be gradually tuned from smooth to rugged. We consider instances with the following settings: the problem size is set to $N = 100$, the number of objectives to $M \in \{2,3,4,5\}$, and the ruggedness to $K \in \{0,1,2,4\}$, that is, from linear to highly rugged landscapes. We generate one instance at random for each combination.

Parameter Setting. For our analysis, we consider three competing algorithms extracted from the MOEA/D–(μ, λ, sps) framework as depicted in Table 1. For the conventional MOEA/D, only one parameter is kept free, that is the population size μ. For MOEA/D–DRA, the sub-problem selection strategy is implemented as described in the original paper [23]. We further consider to experiment MOEA/D–DRA with other λ values. Recall that in the original variant, only $\mu/5$ sub-problems are selected, while including systematically the M boundary weight vectors. For fairness, we follow the same principle when implementing the MOEA/D–RND strategy. Notice that the boundary weight vectors were shown to impact the coordinates of the reference point $z^\star$ used by the scalarizing function [18]. They are then important to consider at each generation. To summarize, for both MOEA/D–DRA and MOEA/D–RND, two parameters are kept free, namely the population size μ and the number of selected sub-problems λ. They are chosen to cover a broad range of values, from very small to relatively very high, namely, $\mu \in \{1,10,50,100,500\}$ and $\lambda \in \{1,2,5,10,25,50,100,150,200,300,400,450,500\}$ such that $\lambda \leqslant \mu$.

The other common parameters are set as follows. The initial weights are generated using the methodology described in [21]. The neighborhood size is set to 20% of the population size: $T = 0.2\,\mu$. Two parents are considered for mating selection, i.e., the parent selection in the neighborhood of a current sub-problem i. The first parent is the current solution x^i, and the second one is selected uniformly at random from $\mathcal{B}_i$. Given that solutions are binary strings, we use a two-point crossover operator and a bit-flip mutation operator where each bit is flipped with a rate of $1/N$. MOEA/D–DRA involves additional parameters which are set following the recommendations from [23].

Performance Evaluation. Given the large number of parameter values (more than 2 000 different configurations), and in order to keep our experiments manageable in a reasonable amount of time, every configuration is executed 10 independent times, for a total of more than 20 000 runs. In order to appreciate the

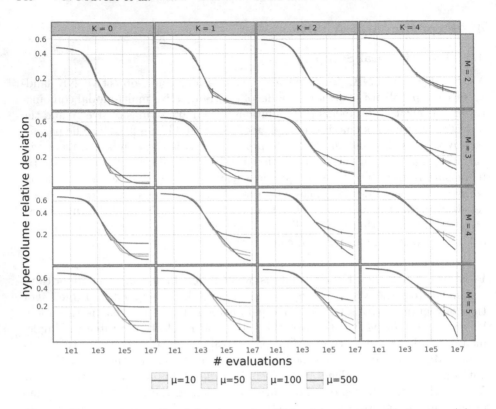

Fig. 1. Convergence profile of the conventional MOEA/D w.r.t. population size (μ).

convergence profile and the anytime behavior of the competing algorithms, we consider different stopping conditions of $\{10^0, 10^1, \ldots, 10^7\}$ calls to the evaluation function. Notice however that due to lack of space, we shall only report our findings on the basis of a representative set of our experimental data.

For performance assessment, we use the hypervolume indicator (hv) [24] to assess the quality of the obtained approximation sets, the reference point being set to the origin. More particularly, we consider the hypervolume relative deviation, computed as $\mathsf{hvrd}(A) = (\mathsf{hv}(R) - \mathsf{hv}(A))/\mathsf{hv}(R)$, where A is the obtained approximation set, and R is the best Pareto front approximation, obtained by aggregating the results over all executions and removing dominated points. As such, a lower value is better. It is important to notice that we consider the external archive, storing all non-dominated points found so far during the search process, for performance assessment. This is particularly important when comparing configurations using different population sizes.

4.2 Impact of the Population Size: sps$_{All}$ with Varying μ Values

We start our analysis by studying the impact of the population size for the conventional MOEA/D, that is MOEA/D–(μ, μ, sps$_{ALL}$) following our terminology. In

Fig. 1, we show the convergence profile using different μ values for the considered instances. Recall that hypervolume is measured on the external archive.

For a fixed budget, a smaller population size allows the search process to focus the computational effort on fewer sub-problems, hence approaching the Pareto front more quickly. By contrast, using a larger population implies more diversified solutions/sub-problems, and hence a better spreading along the Pareto front. This is typically what we observe when a small and a large budget are contrasted. In fact, a larger population size can be outperformed by a smaller one for relatively small budgets, especially when the problem is quite smooth ($K \leqslant 1$) and the number of objectives relatively high ($M \geqslant 3$). Notice also that it is not straightforward to quantify what is meant by a 'small' population, depending on the problem difficulty. For a linear bi-objective problem ($M = 2$, $K = 0$), a particularly small population size of $\mu = 10$ is sufficient to provide a relatively high accuracy. However, for quadratic many-objective problems ($M \geqslant 4$, $K = 1$), a small population size of $\mu = 10$ (resp. $\mu = 50$) is only effective up to a budget of about 10^4 (resp. 10^5) evaluations.

To summarize, it appears that the approximation quality depends both on the problem characteristics and on the available budget. For small budgets, a small population size is to be preferred. However, as the available budget grows, and as the problem difficulty increases in terms of ruggedness and number of objectives, a larger population performs better. These first observations suggest that the anytime behavior of MOEA/D can be improved by more advanced selection strategy, allowing to avoid wasting resources in processing a large number of sub-problems at each iteration, as implied by the conventional sps_{ALL} strategy which iterates over *all* sub-problems. This is further analyzed next.

4.3   Impact of the Sub-problem Selection Strategy

In order to fairly compare the different selection strategies, we analyze the impact of λ, i.e., the number of selected sub-problems, independently for each strategy. It is worth-noticing that *both* the value of λ and the selection strategy impact the probability of selecting a weigh vector. Our results are depicted in Fig. 2 for sps_{DRA} and sps_{RND}, for different budgets and on a representative subset of instances. Other instances are not reported due to space restrictions. The main observation is that the best setting for λ depends on the considered budget, on the instance type, and on the sub-problem selection strategy itself.

Impact of λ on sps_{Rnd}. For the random strategy sps_{RND} (Fig. 2, top), and for smooth problems ($K = 0$), a small λ value is found to perform better for a small budget. As the available budget grows, the λ value providing the best performance starts to increase until it reaches the population size μ. In other words, for small budgets one should select very few sub-problems at each generation, whereas for large budgets selecting all sub-problems at each generation, as done in the standard MOEA/D, appears to be a more reasonable choice. However, this tendency only holds for smooth many-objective problems. When the ruggedness increases, that is when the degree of non-linearity K grows, the effect of λ

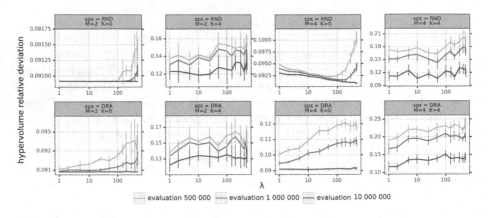

Fig. 2. Quality vs. number of selected sub-problems (λ) w.r.t. budget ($\mu = 500$).

changes. For the highest value of $K = 4$, the smallest value of $\lambda = 1$ still appears to be effective, independently of the available budget. However, the difference with a large λ value is seemingly less pronounced, especially for a relatively large budget, and the effect of λ seems to decay as the ruggedness increases. Notice also that for the 'easiest' problem instance (with $K = 0$ and $M = 2$), it is only for a small budget or for a high λ value that we observe a loss in performance. We attribute this to the fact that, when the problem is harder, search improvements are scarce within all sub-problems, it thus makes no difference to select few or many of them at each generation. By contrast, when the problem is easier, it is enough to select fewer sub-problems, as a small number of improving offspring solutions are likely sufficient to update the population.

Impact of λ on $\mathsf{sps_{Dra}}$. The impact of λ appears to be different when analyzing the $\mathsf{sps_{DRA}}$ strategy (Fig. 2, bottom). In fact, the effect of λ seems relatively uniform, and its optimal setting less sensitive to the available budget and instance type. More precisely, the smallest value of $\lambda = 1$ is always found to perform better, while an increasing λ value leads to a decrease in the overall approximation quality. We attribute this to the adaptive nature of $\mathsf{sps_{DRA}}$, for which the probability of selecting non-interesting sub-problems is smaller for lower λ values. Interestingly, in the original setting of MOEA/D–DRA [23], from which $\mathsf{sps_{DRA}}$ is extracted, the number of selected sub-problems is fixed to $\mu/5$. Not only we found that this setting can be sub-optimal, but it can actually be substantially outperformed by a simple setting of $\lambda = 1$.

$\mathsf{sps_{All}}$ *vs.* $\mathsf{sps_{Dra}}$ *vs.* $\mathsf{sps_{Rnd}}$. Having gained insights about the effect of λ for the different selection strategies, we can fairly analyze their relative performance by using their respective optimal setting for λ. We actually show results with $\lambda = 1$ for both $\mathsf{sps_{DRA}}$ and $\mathsf{sps_{RND}}$. Although this setting was shown to be optimal for $\mathsf{sps_{DRA}}$, it only provides a reasonably good (but sub-optimal) performance in the case of the simple random $\mathsf{sps_{RND}}$ strategy, for which other λ values can be even more efficient. Our results are shown in Fig. 3 for a subset of instances.

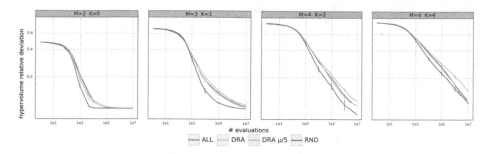

Fig. 3. Convergence profile of MOEA/D–(μ, λ, sps) w.r.t. sub-problem selection strategy ($\mu = 500$; $\lambda = 500$ for $\mathsf{sps}_{\mathrm{ALL}}$, $\lambda \in \{1, \mu/5\}$ for $\mathsf{sps}_{\mathrm{DRA}}$, and $\lambda = 1$ for $\mathsf{sps}_{\mathrm{RND}}$).

Table 2. Ranks and average hvrd value (between brackets, in percentage) obtained by the different sps strategies after $\{10^4, 10^5, 10^6, 10^7\}$ evaluations (a lower value is better). Results for $\mathsf{sps}_{\mathrm{RND}}$ and $\mathsf{sps}_{\mathrm{DRA}}$ are for $\lambda = 1$. For each budget and instance, a rank of c indicates that the corresponding strategy was found to be significantly outperformed by c other strategies w.r.t. a Wilcoxon statistical test at a significance level of 0.05. Ranks in bold correspond to approaches that are not significantly outperformed by any other, and the underlined hvrd value corresponds to the best approach in average.

M	K	10^4 evaluations $\mathsf{sps}_{\mathrm{ALL}}$	$\mathsf{sps}_{\mathrm{DRA}}$	$\mathsf{sps}_{\mathrm{RND}}$	10^5 evaluations $\mathsf{sps}_{\mathrm{ALL}}$	$\mathsf{sps}_{\mathrm{DRA}}$	$\mathsf{sps}_{\mathrm{RND}}$	10^6 evaluations $\mathsf{sps}_{\mathrm{ALL}}$	$\mathsf{sps}_{\mathrm{DRA}}$	$\mathsf{sps}_{\mathrm{RND}}$	10^7 evaluations $\mathsf{sps}_{\mathrm{ALL}}$	$\mathsf{sps}_{\mathrm{DRA}}$	$\mathsf{sps}_{\mathrm{RND}}$
2	0	$2_{(11.2)}$	$1_{(10.5)}$	$\mathbf{0}_{(\underline{09.1})}$	$2_{(09.4)}$	$1_{(09.3)}$	$\mathbf{0}_{(\underline{09.0})}$	$2_{(09.1)}$	$\mathbf{0}_{(09.0)}$	$\mathbf{0}_{(\underline{09.0})}$	$\mathbf{0}_{(09.0)}$	$\mathbf{0}_{(\underline{09.0})}$	$2_{(09.0)}$
	1	$1_{(14.0)}$	$1_{(13.2)}$	$\mathbf{0}_{(\underline{11.4})}$	$1_{(10.9)}$	$1_{(10.5)}$	$\mathbf{0}_{(\underline{09.5})}$	$2_{(09.8)}$	$1_{(09.6)}$	$\mathbf{0}_{(\underline{09.2})}$	$2_{(09.5)}$	$1_{(09.4)}$	$\mathbf{0}_{(\underline{09.2})}$
	2	$1_{(17.2)}$	$\mathbf{0}_{(15.6)}$	$\mathbf{0}_{(\underline{15.4})}$	$1_{(13.0)}$	$\mathbf{0}_{(12.5)}$	$\mathbf{0}_{(\underline{11.7})}$	$1_{(11.3)}$	$\mathbf{0}_{(10.9)}$	$\mathbf{0}_{(\underline{10.6})}$	$\mathbf{0}_{(10.2)}$	$\mathbf{0}_{(10.1)}$	$\mathbf{0}_{(\underline{09.8})}$
	4	$2_{(22.1)}$	$\mathbf{0}_{(19.5)}$	$\mathbf{0}_{(\underline{18.9})}$	$1_{(17.5)}$	$\mathbf{0}_{(\underline{14.9})}$	$\mathbf{0}_{(16.0)}$	$2_{(14.6)}$	$\mathbf{0}_{(13.3)}$	$\mathbf{0}_{(\underline{13.0})}$	$\mathbf{0}_{(13.1)}$	$\mathbf{0}_{(\underline{12.0})}$	$\mathbf{0}_{(12.2)}$
3	0	$1_{(15.1)}$	$1_{(15.0)}$	$\mathbf{0}_{(\underline{10.2})}$	$1_{(11.0)}$	$1_{(10.9)}$	$\mathbf{0}_{(\underline{09.2})}$	$2_{(09.1)}$	$\mathbf{0}_{(09.1)}$	$\mathbf{0}_{(\underline{09.1})}$	$\mathbf{0}_{(09.0)}$	$\mathbf{0}_{(09.0)}$	$2_{(09.1)}$
	1	$1_{(18.4)}$	$1_{(18.4)}$	$\mathbf{0}_{(\underline{14.2})}$	$1_{(13.3)}$	$1_{(13.0)}$	$\mathbf{0}_{(\underline{10.5})}$	$1_{(10.8)}$	$1_{(10.5)}$	$\mathbf{0}_{(\underline{09.4})}$	$1_{(09.5)}$	$1_{(09.5)}$	$\mathbf{0}_{(\underline{09.0})}$
	2	$1_{(24.4)}$	$1_{(23.4)}$	$\mathbf{0}_{(\underline{20.6})}$	$1_{(16.3)}$	$\mathbf{0}_{(16.1)}$	$\mathbf{0}_{(\underline{14.7})}$	$2_{(12.8)}$	$1_{(12.3)}$	$\mathbf{0}_{(\underline{11.1})}$	$1_{(11.7)}$	$1_{(11.1)}$	$\mathbf{0}_{(\underline{09.4})}$
	4	$1_{(31.0)}$	$\mathbf{0}_{(29.9)}$	$\mathbf{0}_{(\underline{28.0})}$	$\mathbf{0}_{(22.7)}$	$\mathbf{0}_{(\underline{20.5})}$	$\mathbf{0}_{(21.4)}$	$1_{(16.7)}$	$\mathbf{0}_{(15.4)}$	$\mathbf{0}_{(15.6)}$	$\mathbf{0}_{(13.6)}$	$\mathbf{0}_{(13.0)}$	$\mathbf{0}_{(\underline{12.2})}$
4	0	$1_{(20.5)}$	$1_{(20.5)}$	$\mathbf{0}_{(\underline{13.2})}$	$1_{(12.9)}$	$1_{(12.8)}$	$\mathbf{0}_{(\underline{09.9})}$	$\mathbf{0}_{(09.4)}$	$\mathbf{0}_{(09.4)}$	$\mathbf{0}_{(\underline{09.3})}$	$\mathbf{0}_{(09.0)}$	$\mathbf{0}_{(09.0)}$	$2_{(09.0)}$
	1	$1_{(25.0)}$	$1_{(24.9)}$	$\mathbf{0}_{(\underline{18.5})}$	$1_{(15.2)}$	$1_{(15.4)}$	$\mathbf{0}_{(\underline{11.8})}$	$1_{(09.9)}$	$1_{(09.8)}$	$\mathbf{0}_{(\underline{08.8})}$	$1_{(08.4)}$	$1_{(08.5)}$	$\mathbf{0}_{(\underline{07.9})}$
	2	$1_{(29.8)}$	$1_{(30.6)}$	$\mathbf{0}_{(\underline{24.7})}$	$1_{(19.5)}$	$1_{(18.7)}$	$\mathbf{0}_{(\underline{16.0})}$	$1_{(12.9)}$	$1_{(12.3)}$	$\mathbf{0}_{(\underline{10.1})}$	$1_{(09.8)}$	$1_{(09.9)}$	$\mathbf{0}_{(\underline{07.9})}$
	4	$1_{(38.0)}$	$1_{(37.1)}$	$\mathbf{0}_{(\underline{32.5})}$	$1_{(26.4)}$	$1_{(25.4)}$	$\mathbf{0}_{(\underline{22.5})}$	$1_{(16.8)}$	$1_{(16.6)}$	$\mathbf{0}_{(\underline{15.1})}$	$\mathbf{0}_{(11.6)}$	$\mathbf{0}_{(11.3)}$	$\mathbf{0}_{(\underline{10.5})}$
5	0	$2_{(26.3)}$	$1_{(25.2)}$	$\mathbf{0}_{(\underline{15.4})}$	$1_{(14.1)}$	$1_{(14.7)}$	$\mathbf{0}_{(\underline{10.2})}$	$\mathbf{0}_{(08.0)}$	$1_{(08.3)}$	$2_{(08.5)}$	$\mathbf{0}_{(\underline{07.4})}$	$\mathbf{0}_{(07.5)}$	$2_{(08.0)}$
	1	$1_{(29.9)}$	$1_{(29.9)}$	$\mathbf{0}_{(\underline{21.5})}$	$1_{(16.5)}$	$1_{(17.1)}$	$\mathbf{0}_{(\underline{13.0})}$	$1_{(08.3)}$	$1_{(08.5)}$	$\mathbf{0}_{(\underline{07.7})}$	$\mathbf{0}_{(\underline{05.9})}$	$\mathbf{0}_{(06.1)}$	$1_{(06.1)}$
	2	$1_{(35.2)}$	$1_{(34.1)}$	$\mathbf{0}_{(\underline{28.1})}$	$1_{(21.4)}$	$\mathbf{0}_{(20.0)}$	$\mathbf{0}_{(\underline{17.6})}$	$\mathbf{0}_{(10.9)}$	$\mathbf{0}_{(10.5)}$	$\mathbf{0}_{(\underline{09.7})}$	$\mathbf{0}_{(06.8)}$	$\mathbf{0}_{(06.2)}$	$\mathbf{0}_{(\underline{05.3})}$
	4	$1_{(41.6)}$	$1_{(40.7)}$	$\mathbf{0}_{(\underline{35.5})}$	$1_{(26.5)}$	$1_{(26.9)}$	$\mathbf{0}_{(\underline{24.1})}$	$\mathbf{0}_{(14.7)}$	$\mathbf{0}_{(15.1)}$	$\mathbf{0}_{(\underline{14.3})}$	$\mathbf{0}_{(\underline{06.0})}$	$\mathbf{0}_{(07.4)}$	$\mathbf{0}_{(07.4)}$

The $\mathsf{sps}_{\mathrm{ALL}}$ strategy, corresponding to the conventional MOEA/D [22], and $\mathsf{sps}_{\mathrm{DRA}}$ with $\lambda = \mu/5$, corresponding to MOEA/D–DRA [23], are also included. We can see that the simple random selection strategy $\mathsf{sps}_{\mathrm{RND}}$ has a substantially better anytime behavior. In other words, selecting a single sub-problem at random is likely to enable identifying a high-quality approximation set more quickly, for a wide range of budgets, and independently of the instance type.

Pushing our analysis further, the only situation where a simple random strategy is outperformed by the conventional MOEA/D or by a MOEA/D–DRA setting

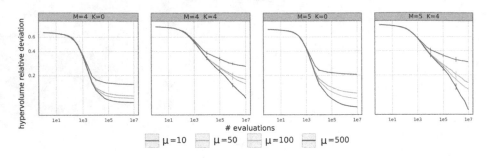

Fig. 4. Convergence profile of MOEA/D–(μ, 1, sps$_{\text{RND}}$) w.r.t. population size (μ).

using an optimal λ value is essentially for the very highest budget (10^7 evaluations) and when the problem is particularly smooth ($K = 0$). This can be more clearly observed in Table 2, where the relative approximation quality of the different strategies are statistically compared for different budgets. Remember however that these results are for $\lambda = 1$, which is shown to be an optimal setting for sps$_{\text{DRA}}$, but not necessarily for sps$_{\text{RND}}$ where higher λ values perform better.

4.4 Robustness of MOEA/D–(μ, λ, sps$_{\text{Rnd}}$) w.r.t. μ and λ

In the previous section, the population size was fixed to the highest value of $\mu = 500$. However, we have shown in Sect. 4.2 that the anytime behavior of the conventional MOEA/D can be relatively sensitive to the setting of μ, in particular for some instance types. Hence, we complement our analysis by studying the sensitivity of the sps$_{\text{RND}}$ strategy, which was found to have the best anytime behavior overall, w.r.t the population size μ. Results for sps$_{\text{RND}}$ with $\lambda = 1$ are reported in Fig. 4. In contrast with the sps$_{\text{ALL}}$ strategy from the conventional MOEA/D reported in Fig. 1, we can clearly see that the anytime behavior underlying sps$_{\text{RND}}$ is much more stable. In fact, the hypervolume increases with μ, independently of the considered budget and instance type. Notice also that when using small μ values, convergence occurs much faster for smooth problems ($K = 0$) compared against rugged ones ($K = 4$). This means that a larger population size μ, combined with a small value of λ, shall be preferred.

From a more general perspective, this observation is quite insightful since it indicates that, by increasing the number of weight vectors, one can allow for a high-level structure of the population, being eventually very large. Notice also that such a data structure can be maintained very efficiently in terms of CPU time complexity, given the scalar nature of MOEA/D. This is to contrast with, e.g., dominance-based EMO algorithms, where maintaining a large population may be computationally intensive, particularly for many-objective problems. Having such an efficient structure, the issue turns out to select some sub-problems from which the (large) population is updated. A random strategy for sub-problem selection is found to work arguably well. However, in order to reach an optimal performance, setting up the number of sub-problems λ might require further configuration issues. Overall, our analysis, reveals that a small λ value, typically ranging from 1 to 10, is recommended for relatively rugged problems, whereas a large value of λ should be preferred for smoother problems.

5   Conclusions and Perspectives

In this paper, we reviewed the design principles of the MOEA/D framework by providing a high-level, but more precise, reformulation taking inspiration from the $(\mu+\lambda)$ scheme from evolutionary computation. We analyzed the role of three design components: the population size (μ), the number of sub-problems selected (and then the number of offspring generated) at each generation (λ), and the strategy used for sub-problems selection (sps). Besides systematically informing about the combined effect of these components on the performance profile of the search process as a function of problem difficulty in terms of ruggedness and objective space dimension, our analysis opens new challenging questions on the design and practice of decomposition-based EMO algorithms.

Although we are now able to derive a parameter setting recommendation according to the general properties of the problem at hand, such properties might not always be known beforehand by the practitioner, and other properties might be considered as well. For instance, one obvious perspective would be to extend our analysis to the continuous domain. More importantly, an interesting research line would be to infer the induced landscape properties in order to learn the 'best' parameter setting, either off-line or on-line; i.e. before or during the search process. This would not only avoid the need of additional algorithm configuration (tuning) efforts, but it could also lead to an even better anytime behavior. One might for instance consider an adaptive setting where the values of μ, λ, and sps are adjusted according to the search behavior observed over different generations. Similarly, we believe that considering a problem where the objectives expose some degree of heterogeneity, e.g., in terms of solving difficulty, is worth investigating. In such a scenario, the design of an accurate sps strategy is certainly a key issue. More generally, we advocate for a more systematic analysis of such considerations for improving our fundamental understanding of the design issues behind MOEA/D and EMO algorithms in general, of the key differences between EMO algorithm classes, and of their success in solving challenging multi- and many-objective optimization problems.

Acknowledgments. This work was supported by the French national research agency (ANR-16-CE23-0013-01) and the Research Grants Council of Hong Kong (RGC Project No. A-CityU101/16).

References

1. Aghabeig, M., Jaszkiewicz, A.: Experimental analysis of design elements of scalarizing function-based multiobjective evolutionary algorithms. Soft. Comput. **23**(21), 10769–10780 (2018). https://doi.org/10.1007/s00500-018-3631-x

2. Beume, N., Naujoks, B., Emmerich, M.: SMS-EMOA: multiobjective selection based on dominated hypervolume. Eur. J. Oper. Res. **181**(3), 1653–1669 (2007)

3. Cai, X., Li, Y., Fan, Z., Zhang, Q.: An external archive guided multiobjective evolutionary algorithm based on decomposition for combinatorial optimization. IEEE Trans. Evolut. Comput. **19**(4), 508–523 (2015)

4. Chiang, T., Lai, Y.: MOEA/D-AMS: improving MOEA/D by an adaptive mating selection mechanism. In: CEC 2011, pp. 1473–1480 (2011)

5. Corus, D., Oliveto, P.S.: Standard steady state genetic algorithms can Hillclimb faster than mutation-only evolutionary algorithms. IEEE Trans. Evol. Comput. **22**(5), 720–732 (2018)

6. Črepinšek, M., Liu, S.H., Mernik, M.: Exploration and exploitation in evolutionary algorithms: a survey. ACM Comput. Surv. **45**(3), 1–33 (2013)

7. Deb, K.: Multi-objective Optimization Using Evolutionary Algorithms. Wiley, Hoboken (2001)

8. Glasmachers, T., Naujoks, B., Rudolph, G.: Start small, grow big? Saving multi-objective function evaluations. In: Bartz-Beielstein, T., Branke, J., Filipič, B., Smith, J. (eds.) PPSN 2014. LNCS, vol. 8672, pp. 579–588. Springer, Cham (2014). https://doi.org/10.1007/978-3-319-10762-2_57

9. Ishibuchi, H., Imada, R., Masuyama, N., Nojima, Y.: Two-layered weight vector specification in decomposition-based multi-objective algorithms for many-objective optimization problems. In: CEC, pp. 2434–2441 (2019)

10. Lavinas, Y., Aranha, C., Ladeira, M.: Improving resource allocation in MOEA/D with decision-space diversity metrics. In: Martín-Vide, C., Pond, G., Vega-Rodríguez, M.A. (eds.) TPNC 2019. LNCS, vol. 11934, pp. 134–146. Springer, Cham (2019). https://doi.org/10.1007/978-3-030-34500-6_9

11. Li, H., Zhang, Q.: Multiobjective optimization problems with complicated Pareto sets, MOEA/D and NSGA-II. IEEE Trans. Evol. Comput. **13**(2), 284–302 (2009)

12. Li, K., Zhang, Q., Kwong, S., Li, M., Wang, R.: Stable matching-based selection in evolutionary multiobjective optimization. IEEE TEC **18**(6), 909–923 (2014)

13. Marquet, G., Derbel, B., Liefooghe, A., Talbi, E.-G.: Shake them all! Rethinking selection and replacement in MOEA/D. In: Bartz-Beielstein, T., Branke, J., Filipič, B., Smith, J. (eds.) PPSN 2014. LNCS, vol. 8672, pp. 641–651. Springer, Cham (2014). https://doi.org/10.1007/978-3-319-10762-2_63

14. Schumer, M., Steiglitz, K.: Adaptive step size random search. IEEE Trans. Autom. Control **13**(3), 270–276 (1968)

15. Tanabe, R., Ishibuchi, H.: An analysis of control parameters of MOEA/D under two different optimization scenarios. Appl. Soft Comput. **70**, 22–40 (2018)

16. Trivedi, A., Srinivasan, D., Sanyal, K., Ghosh, A.: A survey of multiobjective evolutionary algorithms based on decomposition. IEEE TEC **21**(3), 440–462 (2017)

17. Verel, S., Liefooghe, A., Jourdan, L., Dhaenens, C.: On the structure of multiobjective combinatorial search space: MNK-Landscapes with correlated objectives. Eur. J. Oper. Res. **227**(2), 331–342 (2013)

18. Wang, P., et al.: A new resource allocation strategy based on the relationship between subproblems for MOEA/D. Inf. Sci. **501**, 337–362 (2019)

19. Wang, Z., Zhang, Q., Zhou, A., Gong, M., Jiao, L.: Adaptive replacement strategies for MOEA/D. IEEE Trans. Cybern. **46**(2), 474–486 (2016)

20. Witt, C.: Population size versus runtime of a simple evolutionary algorithm. Theor. Comput. Sci. **403**(1), 104–120 (2008)

21. Zapotecas-Martínez, S., Aguirre, H., Tanaka, K., Coello, C.: On the low-discrepancy sequences and their use in MOEA/D for high-dimensional objective

spaces. In: Congress on Evolutionary Computation (CEC 2015), pp. 2835–2842 (2015)

22. Zhang, Q., Li, H.: MOEA/D: a multiobjective evolutionary algorithm based on decomposition. IEEE Trans. Evol. Comput. **11**(6), 712–731 (2007)

23. Zhou, A., Zhang, Q.: Are all the subproblems equally important? Resource allocation in decomposition-based multiobjective evolutionary algorithms. IEEE Trans. Evol. Comput. **20**(1), 52–64 (2016)

24. Zitzler, E., Thiele, L., Laumanns, M., Fonseca, C.M., Grunert da Fonseca, V.: Performance assessment of multiobjective optimizers: an analysis and review. IEEE Trans. Evol. Comput. **7**(2), 117–132 (2003)

A Grouping Genetic Algorithm for Multi Depot Pickup and Delivery Problems with Time Windows and Heterogeneous Vehicle Fleets

Cornelius Rüther[✉] and Julia Rieck

University of Hildesheim, Universitätsplatz 1, 31141 Hildesheim, Germany
{ruether,rieck}@bwl.uni-hildesheim.de

Abstract. The Multi Depot Pickup and Delivery Problem with Time Windows and Heterogeneous Vehicle Fleets is a Rich Vehicle Routing Problem as it combines many real-world problems and is therefore relevant to practice. In this paper a new mathematical two-index model formulation for the MDPDPTWHV is developed as well as a Grouping Genetic Algorithm (GGA), which features a grouping-oriented individual representation. Therefore, each chromosome contains only the assignment of requests to vehicles, i.e., no information about the customer sequence is included. In order to compare different variants of the GGA to each other as well as the best one to solutions calculated by Cplex, 120 MDPDPTWHV datasets are created through a generator implemented by the authors. In a benchmark study, it can be shown that the way in which population management is performed is important to enhance the solution quality of the GGA. On average, the best GGA variant is 2.43% worse than the best known solution.

Keywords: Multi-Depot Pickups and Deliveries · Combinatorial optimization model · Rich Vehicle Routing

1 Problem Identification

In Germany, almost all of the top 10 transportation companies (with respect to their revenue) operate in the Less-Than-Truckload (LTL) sector [23] in which several transportation requests are transported together in one truck. Usually, the freight of each request has to be forwarded from an origin (pickup customer) to a destination (delivery customer). At each customer location, the corresponding goods have to be loaded by employees. As this is a time-sensitive process due to the availability of employees, loading ramps, and equipment, it is important to consider loading time windows in order to reduce waiting times [8,9]. Since different types of goods are transported, e.g., palletized goods, but also lattice boxes, cable reels or steel rods, LTL carriers have to maintain different vehicle types with regard to capacity, speed or even pollution emission. Large carriers in

© Springer Nature Switzerland AG 2020
L. Paquete and C. Zarges (Eds.): EvoCOP 2020, LNCS 12102, pp. 148–163, 2020.
https://doi.org/10.1007/978-3-030-43680-3_10

the LTL market typically have several locations, so that a multi depot problem must be taken into account. At each depot, vehicles are available which start and end their routes at this depot.

The described application appears not only with large carriers. In the event that small LTL companies operate in cooperations, a central collaborative pickup and delivery problem consisting of data from all carriers needs to be considered. Due to the fact that about 60% of carriers have a small vehicle fleet of at most 10 vehicles [5], these carriers usually have only one depot. As a concept to retain their competitive ability, a possible and clever strategy for these small and medium-sized carriers is to affiliate in coalitions, in order to collaborate in transportation planning. Already 20% of the top 25 carriers on the LTL market (taking the revenue into account) were organized in coalitions of small companies in 2015 [11]. It is known from [13] that there are many contributions with centrally organized cooperation concepts, i.e., a central entity optimizes the transportation plans over all participants of the coalition (multi depot problem). For this purpose, this entitiy assigns all requests cost-efficiently to the coalition partners, while some of the requests may have been acquired by partners other than the assigned partners. By this request exchange, the coalition profit is gained and is allocated by the central entity to all participants.

The applications result in a practical problem that occurs at both large carriers and small cooperating carriers. In particular, it is a special Pickup and Delivery Problem (PDP), where requests with a positive quantity have to be transported from origins to destinations on the same vehicle or route, respectively. Each vehicle starts and ends at a certain depot, while several depots are taken into account (Multi Depot). The vehicles themselves may be heterogeneous regarding, e.g., capacity (HV). In order to address the bottleneck of employees and equipment for loading activities, (hard) time windows (TW) are taken into account in which the service must be started. Hence, the proposed LTL problem is modelled as a Multi Depot Pickup and Delivery Problem with Time Windows and Heterogeneous Vehicle Fleets (MDPDPTWHV), which is a problem of the Rich Vehicle Routing Problem (RVRP) class, since a variety of restrictions relevant in practice are integrated [4,16] (cf. model formulation in Sect. 3).

The outline of the paper is as follows: We first give a brief overview of the related literature in Sect. 2. A two-index formulation of the proposed MDPDPTWHV is presented in Sect. 3. A detailed description of the solution approach, which is based on the group-oriented Genetic Algorithm presented in [10,19], can be found in Sect. 4. To evaluate this Genetic Algorithm (GA), we consider a benchmark test set with 120 self-generated MDPDPTWHV instances, where the structure of the data is similar to the well known Li and Lim PDPTW datasets [18]. Computational results are presented in detail in Sect. 5. Finally, we close the paper in Sect. 6 with a short discussion.

2   Related Work

In this section, we give an overview of contributions that deal with problems of the RVRP class. We focus on articles that either consider a problem that is similar to our routing problem or use a solution method that is similar to our approach. In addition, different variants of RVRPs with appropriate solution methods can be found in [4, 16, 27].

The PDP with time windows is addressed by Ropke and Pisinger [21], who developed a large neighborhood search. In their approach, the authors apply several competing insertion and removal heuristics (e.g., shaw removal, regret-2 insertion), which are chosen by a roulette wheel selection parameterized by the historical performance of the heuristics. The effectiveness of the approach is shown using benchmark instances from the literature on the one hand and new randomly generated instances containing multiple depots on the other hand. Li et al. [17] propose a PDPTW with profits and reserved requests, which is a routing problem with selective requests for carrier collaboration, i.e., some requests can be reserved by the owning carriers. An adaptive large neighborhood search is developed to solve the problem. The results obtained by the solution method are compared to the results of an exact solver. Goel and Gruhn [14] also present an approach based on large neighborhood search which is able to address practical constraints of multiple pickup and delivery problems, such as time windows, heterogeneous vehicles, and also restrictions that refer to order-vehicle-assignments. In the paper of Irnich [15], a special kind of multi depot pickup and delivery problem is considered, in which all requests has to be transported via a central location (called hub). The routes before and beyond the hubs are typically short, thus the routes can be easily enumerated. The author has derived lower bounds for the addressed problem as well as strategies how to integrate the model into a column generation or branch-and-price approach.

Dondo and Cerda [7] consider a multi depot vehicle routing problem (VRP) with time windows and heterogeneous vehicle fleets. In contrast to the PDP, a vehicle routing problem has the advantage, that the place of pickup is always a depot. The authors take vehicles into account that are not permanently assigned to one depot, i.e., open routes can be performed. They developed a cluster-based optimization approach to solve the problem. This procedure assigns cost-efficient customer clusters to a vehicle and considers the sequencing of each cluster by solving a mixed-integer model. In the paper of Sombuntham and Kachitvichyanukul [25], a multi depot VRP variant with simultaneous pickup and delivery requirements at customer nodes is introduced. The problem has the characteristic that goods picked up at a customer location do not necessarily have to be transported to a depot and places of delivery can be different from depots. The authors implement a particle swarm optimization technique with learning structures in order to solve the described problem with heterogeneous vehicles. Bettinelli et al. [3] consider a problem slightly different to the problem in Sect. 1, where soft time windows are taken into account, i.e., the time windows need not be yield, but a penalty price is charged for violating the time window. A branch-and-price algorithm is developed to analyze the impact of time window management on the overall distribution costs.

Alaia et al. [1] present a genetic algorithm for multi depot PDPTW, where all vehicles have the same capacity and each vehicle starts and ends its route at the same depot. The genetic algorithm works with a solution representation that saves the ordering of the customer nodes which have to be visited (path representation). A detailed description is given of how the initial population is created. In order to hold the problem constraints, several repair procedures are necessary. The PDPTW is also taken into account by Pankratz [19]. In order to solve the problem, a Grouping Genetic Algorithm (GGA) is used. Since the problem contains precedence constraints (pickup before delivery), a typical chromosome representation for VRPs leads to frequently occurring infeasible solutions. Moreover, the crossover and mutation operators have to be very complex to handle the problem (cf. [1]). Therefore, the GGA saves in the chromosomes which vehicles serve which requests. The efficiency of the algorithm is proven by using the Li and Lim PDPTW datasets [24]. A comparison of the GGA to the tabu search presented in the paper of Li and Lim [18] shows that the GGA performs best.

3   A Two Index Formulation for the MDPDPTWHV

In order to get an impression of the problem structure for the Pickup and Delivery Problem at hand (cf. Sect. 1), we introduce a new two-index model formulation. The main advantage of the model in comparison to a typical three-index formulation (cf. [6,22]) is the significant reduction of decision variables. This reduction enables to calculate solutions with exact solvers (such as Cplex) even for large, complex problem instances (cf. performance results in Sect. 5.2). In order to take time windows or heterogeneous vehicles into account, a vehicle index must be introduced. Usually, this results in three-index decision variables x_{ij}^v that indicate whether a vehicle v uses the arc from node i to j in the underlying directed transportation graph. In our advantageous two-index model, we now split up the required decision in two binary decision variables: x_{ij} to decide if an arc (i, j) is used or not, and y_{iv} to determine if a customer i is visited by vehicle v. In detail, the parameters and variables are listed in Table 1.

$$\min \quad \sum_{i,j \in \mathcal{V}} c_{ij} x_{ij} + \sum_{v \in \mathcal{K}} \sum_{d \in \mathcal{D}_s} f_v y_{dv} \tag{1}$$

$$\text{s.t.} \quad \sum_{j \in \mathcal{N} \cup \mathcal{D}_e} x_{ij} = 1 \qquad \forall i \in \mathcal{N} \tag{2}$$

$$\sum_{i \in \mathcal{N} \cup \mathcal{D}_s} x_{ij} = 1 \qquad \forall j \in \mathcal{N} \tag{3}$$

$$\sum_{j \in \mathcal{N}_p} x_{dj} \leq k \qquad \forall d \in \mathcal{D}_s \tag{4}$$

$$\sum_{j \in \mathcal{N}_p} x_{dj} - \sum_{v \in \mathcal{K}} y_{dv} = 0 \qquad \forall d \in \mathcal{D}_s \tag{5}$$

$$\sum_{i \in \mathcal{N}_d} x_{id} - \sum_{v \in \mathcal{K}} y_{dv} = 0 \qquad \forall d \in \mathcal{D}_e \tag{6}$$

$$\sum_{v \in \mathcal{K}} y_{iv} = 1 \qquad \forall i \in \mathcal{N}_p \tag{7}$$

Table 1. Variables and parameters (alphabetically ordered)

Indices and sets:	
$d \in \mathcal{D}$	Set of depots, where each depot is modelled as start depot d and as end depot $2n + \delta + d$; δ is the number of depots $\Rightarrow$ set of start depots $\mathcal{D}_s = \{0, \ldots, \delta - 1\}$, set of end depots $\mathcal{D}_e = \{2n + \delta, \ldots, 2n + 2\delta - 1\}$
$i, j \in \mathcal{N}$	Set of customer nodes, $\mathcal{N} = \{\delta, \ldots, 2n + \delta - 1\}$
$i, j \in \mathcal{N}_p$	Set of pickup locations, $\mathcal{N}_p = \{\delta, \ldots, n + \delta - 1\}$
$i, j \in \mathcal{N}_d$	Set of delivery locations, $\mathcal{N}_d = \{n + \delta, \ldots, 2n + \delta - 1\}$
$i, j \in \mathcal{V}$	Set of all nodes, $\mathcal{V} = \mathcal{N} \cup \mathcal{D}$
$r \in \mathcal{R}$	Set of requests, $\mathcal{R} = \{1, \ldots, n\}$, where each request consists of a pickup location $i \in \mathcal{N}_p$ and delivery location $j \in \mathcal{N}_d$, $j = i + n$
$v \in \mathcal{K}$	Set of (heterogeneous) vehicles, $\mathcal{K} = \{1, \ldots, k\}$

Parameters:	
$[a_i, b_i]$	Time window at node $i \in \mathcal{V}$ in which the service has to start
$c_{ij}(t_{ij})$	Variable costs (travel time) for travelling from i to $j \in \mathcal{V}$
d_i	Delivery demand at customer $i \in \mathcal{N}$, where $d_i > 0$, $\forall i \in \mathcal{N}_d$, and $d_i = 0$, $\forall i \in \mathcal{N}_p$; please note that $d_{i+n} = p_i$, $\forall i \in \mathcal{N}_p$ holds
f_v	Fixed costs for using a vehicle $v \in \mathcal{K}$
κ_v	Capacity of vehicle $v \in \mathcal{K}$
κ_{max}	Maximum capacity of all vehicles
l	Load factor for loading the vehicles measured in time unit per quantity unit
l_{const}	Additional loading time to model preparations before service
M_κ, M_T	Big M's for linearization of time window and capacity constraints
p_i	Pickup demand at customer $i \in \mathcal{N}$, where $p_i > 0$, $\forall i \in \mathcal{N}_p$, and $p_i = 0$, $\forall i \in \mathcal{N}_d$

Decision variables:	
L_i	Current load of the visiting vehicle after serving node $i \in \mathcal{V}$
S_i	Service-time at node $i \in \mathcal{V}$ (depending on the demand $d_i + p_i$)
T_i	Beginning of the service at node $i \in \mathcal{V}$
x_{ij}	Binary variable which indicates if a vehicle uses the arc between nodes i and $j \in \mathcal{V}$
y_{iv}	Binary variable which specifies if vehicle $v \in \mathcal{K}$ serves node $i \in \mathcal{V}$

$$y_{iv} - y_{jv} \leq 1 - x_{ij} \qquad \forall i, j \in \mathcal{N}, v \in \mathcal{K} \qquad (8)$$

$$x_{di} + x_{dj} + y_{iv} + y_{jv} \leq 3 \qquad \forall i, j \in \mathcal{N}_p, v \in \mathcal{K}, d \in \mathcal{D}_s \quad (9)$$

$$x_{id} + x_{jd} + y_{iv} + y_{jv} \leq 3 \qquad \forall i, j \in \mathcal{N}_d, v \in \mathcal{K}, d \in \mathcal{D}_e \ (10)$$

$$\sum_{i \in \mathcal{N}_p} y_{iv} \leq n \sum_{d \in \mathcal{D}_s} y_{dv} \qquad \forall v \in \mathcal{K} \qquad (11)$$

$$\sum_{d \in \mathcal{D}_s} y_{dv} \leq 1 \qquad \forall v \in \mathcal{K} \tag{12}$$

$$y_{iv} - y_{i+n,v} = 0 \qquad \forall v \in \mathcal{K}, i \in \mathcal{N}_p \tag{13}$$

$$y_{d_s v} - y_{d_e v} = 0 \qquad \forall v \in \mathcal{K}, d_s, d_e \in \mathcal{D} \tag{14}$$

$$T_i + S_i + t_{ij} - T_j \leq M_T(1 - x_{ij}) \qquad \forall i,j \in \mathcal{V}, i \notin \mathcal{D}_e, j \notin \mathcal{D}_s \tag{15}$$

$$a_i \leq T_i \leq b_i \qquad \forall i \in \mathcal{V} \tag{16}$$

$$T_i + S_i + t_{i,i+n} \leq T_{i+n} \qquad \forall i \in \mathcal{N}_p \tag{17}$$

$$l \cdot (p_i + d_i) + l_{\text{const}} = S_i \qquad \forall i \in \mathcal{V} \tag{18}$$

$$L_i - d_j + p_j - L_j \leq \kappa_{\max}(1 - x_{ij}) \qquad \forall i,j \in \mathcal{V}, i \notin \mathcal{D}_e, j \notin \mathcal{D}_s \tag{19}$$

$$L_i - \kappa_v \leq M_\kappa(1 - y_{iv}) \qquad \forall i \in \mathcal{V}, v \in \mathcal{K} \tag{20}$$

$$x_{ij}, y_{iv} \in \{0,1\} \qquad \forall i,j \in \mathcal{V}, v \in \mathcal{K} \tag{21}$$

$$L_i, S_i, T_i \in \mathbb{R}^{\geq 0} \qquad \forall i \in \mathcal{V} \tag{22}$$

Objective function (1) considers variable costs regarding the length of all routes and the fixed costs of all used vehicles that are to be minimized. Constraints (2) and (3) ensure that each customer is visited exactly once. Moreover, the number of used vehicles should not exceed the fleet size k at the corresponding depot (4). Please note that the first customer in a route has to be a pickup and the last customer a delivery node as well as a feasible route can only start at a *start depot* and end at the associated *end depot*. Constraints (5) and (6) guarantee that an arc starting from a depot (and ending in a depot) is associated with a vehicle. Each pickup location must be visited by exactly one vehicle (7). If customer j directly follows on i, both of them should be served by the same vehicle v (8). Constraints (9) and (10) make sure that each starting and ending arc of a route will be assigned to a unique vehicle. With condition (11) it is ensured that a vehicle $v \in \mathcal{K}$ may serve customers, if and only if it starts from the depot. In addition, a maximum number of one vehicle can be assigned to one start depot (12). Constraints (13) ensure that the pickup and delivery location of a specific request are carried on the same route. Moreover, (14) guarentee that each starting vehicle ends in the corresponding depot. If customer j directly follows on i, the beginning of service at j is not allowed to start before the end of service at i plus the travel time between i and j (15). The service start time must be in the given time window $[a_i, b_i]$ of customer i (16). The precedence relationship between pickup and delivery nodes is modeled using inequalities (17), where the beginning of services at node i and $i + n$ are considered. The constraints (18) set the service time, which depends on the demand $p_i + d_i$ at customer i and a constant loading time, e.g., for the preparation of loading. Restrictions (19) ensure that the load L_i after visiting customer i will be updated correctly by subtracting the delivered and adding the picked up load. Furthermore, (20) guarentee that the capacity of the vehicle used to serve customer i will not be exceeded. Note that we can choose $M_T = \xi - \eta$ and $M_\kappa = \kappa_{\max} - \kappa_{\min}$, whereas $[\eta, \xi]$ is the time window of the depot and $\kappa_{\max}, \kappa_{\min}$ are the maximum and minimum capacity of the vehicles. Finally, we have to restrict transportation variables to be binary (21) as well as time and load variables to be non-negative, real numbers (22).

It can be shown that a three-index formulation of the MDPDPTWHV contains $(2n + 2d)^2(k - 1) - (2n + 2d)k$ more decision variables for $k > 1$. Therefore, the two-index formulation has quadratically less decision variables for all problem instances with more than one vehicle, which speeds up exact solvers.

4   A Grouping Genetic Algorithm

In the literature, GAs are known to give good results especially for RVRP, such as the Grouping Genetic Algorithm for the PDPTW introduced by Pankratz in [19]. It can hence be assumed that a genetic algorithm promises a good solution quality for the presented problem. Thus, we used the GGA as a basis for solving the MDPDPTWHV and extended the procedure regarding, in particular, crossover and mutation operators. Due to a fixed vehicle depot assignment and our solution representation, the GGA manages the multi depot aspect automatically. An overview of the procedure of the GGA and the implemented approaches in each step is given in Fig. 1. Please note that all terms and concepts of genetic algorithms can be read in Eiben and Smith [12].

In general, the presented GGA proceeds as follows: Initially, a population of n_{pop} individuals is determined. We assume that each individual has a fitness value that is equal to the sum of variable and fixed costs of the vehicle routing solution as described in objective function (1), which has to be minimized in the proposed GA variant. Each individual is computed by using two double insertion heuristics, where both of them are sequential insertion heuristics that create routes one after another by inserting customers. In order to enhance the diversity of the initial population, we generate 25% of the individuals with *Best Insertion* that starts the route with one request and inserts the best corresponding request in each iteration, and 75% with *Random* which chooses a random request in every iteration and inserts it at the best position within the current route. A high level of diversity is achieved, as we do not allow duplicates within the population. For this purpose, a comparison of the objective values, the numbers of vehicles used, and the fixed costs is carried out, since each pair of the three attributes is not unique for comparing two individuals.

As long as a termination criterium has not been met, two parents p_1, p_2 for generating offsprings are selected. The termination criterium may depend on the number of generations γ_{max} or on individuals generated (generally η_{max} or without fitness improvement $\tilde{\eta}_{max}$). With a probability of p_{cross}, one crossover operator is applied to p_1, p_2 in order to create two children c_1, c_2. Moreover, each child is modified by using one of the mutation operations with probability p_{mut} and so two mutated children $\tilde{c}_1$ and $\tilde{c}_2$ are generated. In case no operator is applied, the created offsprings are just clones of their ancestors. Finally, the best individual regarding its fitness is selected as result. Please note that $n_{pop}, \gamma_{max}, \eta_{max}, \tilde{\eta}_{max}, p_{cross}, p_{mut}$ are parameters that have to be set manually.

In general, the most critical part of a GA is to find a good solution representation or genotype encoding, respectively, such that crossover and mutation

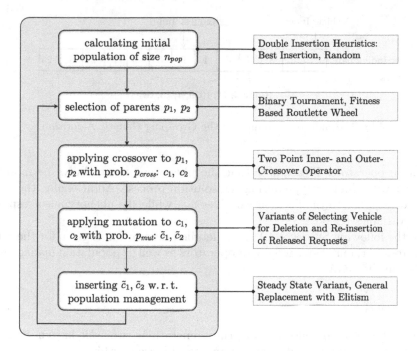

Fig. 1. Grouping Genetic Algorithm Framework (left) and its implemented approaches (right)

operators do not have to be too complex as well as the decoding from genotype to phenotype is not too time expensive. In particular, for pickup and delivery problems a genotype encoding that saves the ordering of customers within the chromosome is not practical, since the operators then have to take a lot of constraints into account, such as precedence relationships, and their verification can be very time-consuming. This is why our approach, as well as the one from Pankratz in [19], is implemented as a Grouping GA, which has been introduced by Falkenauer in [10] and groups the pickup and delivery customers to their request index. A genotype encoding then saves the assignment of a request to the executing vehicle and can be represented by a vector of integers (cf. Fig. 2). The corresponding phenotype, i.e., the associated routes like route 6 in Fig. 2, is calculated through double insertion heuristics as described above.

As demonstrated in Fig. 2, vehicles are represented by negative index values and all subsequent positive integers characterize the requests served by the corresponding vehicle. This encoding implies that for each vehicle the route (e.g., route 6 in Fig. 2) has to be calculated, whenever the genotype decoding is necessary, e.g., for fitness value determination. Since this is a time-consuming procedure, which is with respect to the applied heuristic not necessarily deterministic, we also store the corresponding phenotype of an individual (cf. [19]), i.e., the vehicle routing solution. In this way, good features of an individual are less likely to be removed from the phenotype, when applying the crossover or

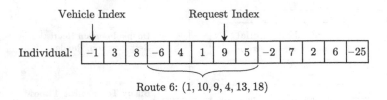

Fig. 2. Genotype encoding for the Grouping Genetic Algorithm

mutation operators. Please note that the phenotype must always be modified together with the genotype during the solution process. Additionally, the genotype corresponds to the decision variables y_{iv}, while the phenotype corresponds to x_{ij} variables (cf. Sect. 3).

In the following Subsects. 4.1–4.5 a detailed description is given for the selection, crossover, mutation, and repair operators as well as population management applied in the GGA.

4.1   Selection

In order to select parents for offspring generation, two well known selection techniques (cf. [20]) are implemented and can be used within our GGA.

Fitness Proportional. Two individuals are chosen with regard to their fitness value. To do so, each slot on a roulette wheel represents an individual with its slotsize determined by the fitness, hence individuals with better fitness are chosen with a higher probability.

Binary Tournament. Two individuals are chosen randomly without duplicates from the population and the best individual is chosen as first parent. The second parent is selected analogously.

4.2   Crossover Operator

The implemented crossover as in [19] is based on the group-oriented crossover variant introduced by Falkenauer [10]. A simplified illustration of how the crossover works for a representation with nine requests distributed over 25 possible vehicles, of which three/four are used, is given in Fig. 3. Firstly, we choose randomly two crossover points for the first parent p_1 by selecting two genes and finding the nearest vehicle indices to the left and right of these selected genes, i.e., there are always only complete vehicles crossed. In contrast to Pankratz [19], we then cross (or transfer) the genes in between the two crossing points (inner genes) in half of the cases and otherwise the genes outside of those (outer genes). In doing so, we investigate the solution space in a broader way. To insert the chosen sequence of genes, an insertion point (that is also a vehicle index) in the second parent p_2 is selected randomly, at which the genes to be crossed are placed. Since this replacement might lead to infeasibility with respect to the

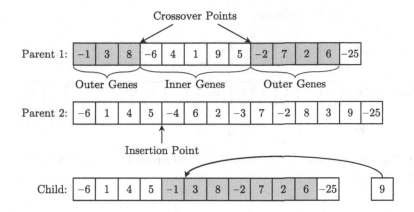

Fig. 3. Crossover operator when transferring genes from parent 1 to parent 2

vehicles used or the served requests, we need to investigate these aspects to fix infeasibilities. If a vehicle to be inserted is already present in parent p_2, we delete this vehicle from parent p_2 and insert the newly selected one from p_1. Due to the deletion of doubled vehicles, it is possible that there are unassigned requests. These will be re-inserted by using the repair operator explained in Subsect. 4.5. The second child is generated analogously by switching notation of p_1 and p_2. While modifying the genotype, the phenotype will be changed accordingly.

4.3   Mutation Operator

The mutation is done through selecting a vehicle index and deleting the corresponding vehicle from the genotype and phenotype. All requests served by the erased vehicle have to be re-inserted, which is done by the repair operator (cf. Subsect. 4.5), such that the solution yields all constraints, i.e., possibly a new vehicle has to be introduced. Contrary to Pankratz [19], we use the following vehicle selection mechanisms to tackle the problem of heterogeneous vehicles. A simplified scheme of the mutation operator is given in Fig. 4.

Costs/Number-of-Request Ratio. For each vehicle, the ratio of route costs and the number of served requests is calculated and used to generate a roulette wheel. By spinning this roulette wheel, a vehicle index is selected, i.e., expensive routes with a small number of requests are preferred for removal. This variant is applied in 40% of the cases.

Number of Requests. The vehicle with minimum number of requests is used for deletion. This variant is selected with a probability of 0.4.

Random Vehicle Index. Here, in 10% of the cases a random vehicle index from the genotype is selected, so all vehicles are equally likely to be chosen.

Random Genotype Position. This variant chooses a random position within the genotype and selects the corresponding vehicle index. In doing so, vehicles with many requests are preferred. The variant is applied with probability of 0.1.

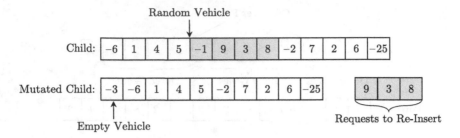

Fig. 4. Mutation operator and mutated child to be repaired (using an additional empty vehicle if needed)

We also extend the mutation operator by applying a so-called swap operator with a certain probability, after one of the mutation variants is applied. This operator tries to swap used and free vehicles to minimize the sum of fixed costs. In order to keep the search broad within the solution space, the choice of vehicles to be swapped is controlled by a roulette wheel.

4.4 Population Management

Various variants can be used for population management. As we know from [26], a steady-state population management can produce better results than a general replacement while being high performant, which it is shown in [19]. Contrary to the PDPTW addressed by Pankratz, the MDPDPTWHV considered in this paper has a larger and more complex solution space, i.e., in a steady-state variant the intensification may be too high and the diversification too low to search the solution space broadly enough. Thus, we have implemented a variant of general replacement with elitism in this framework. In our general replacement variant, a mating pool of 1.5 of the generation size is filled with generated children, whereas duplicates are prevented. To preserve the best solutions from the last generation, elitism is applied by copying the best individuals (5% of generation size) into the next generation. Then, the best 5% of the individuals from the mating pool are selected and the rest of the next generation is filled with randomly chosen individuals from the mating pool. The GGA provides also the option of using a steady-state variant, in which the generated children $\tilde{c}_1, \tilde{c}_2$ replace the two worst individuals in the population, if no duplicates are produced.

4.5 Repair Operator

When applying the crossover or mutation, respectively, we often have to fill the chromosome with unattended/non-served requests. In order to re-insert these requests, we use a double insertion heuristic, that is calculating the best insertion positions for each request and places the best one at its computed position, i.e., the heuristic is greedy. This approach is quite similar to the best (double) insertion heuristic, which is used for generating the population. However, the proposed repair operator is a parallel insertion, i.e., the possible insertion positions are evaluated for all vehicles being used. If there are requests that cannot be served by the available vehicles, a new (empty) vehicle has to be introduced.

5 Evaluation

In this section, the results of four Grouping Genetic Algorithm variants are presented, which includes a variant that is similar to the one from Pankratz [19]. At first, it is described how the benchmark datasets are created (cf. Subsect. 5.1). Then, the results for all GGA variants are presented. Furthermore, the best and most stable GGA variant is compared to the solution calculated by Cplex (cf. Subsect. 5.2).

5.1 Data Generation

In order to create datasets for GGA evaluation, a generator implemented by the authors for PDPTWHV datasets is used. A multi-depot environment is created during the generation process by creating four single depot PDPTWHV instances and layering them on the top of each other, i.e., the instances have four depots with up to 80 requests. To make the generated datasets comparable with Li and Lim PDPTW datasets [18], different parameters (e.g., number of requests, cluster size, time window structures, vehicle capacity etc.) may be set. All parameters used for generating instances, which can be downloaded from [2], are given in Table 2. In order to create 60 datasets, 12 variants of all possible parameter combinations are applied on five generate runs. In doing so, instances with the combination, e.g., (s3, tw2, v3) correspond to the randomly created datasets with large time windows and big vehicles, which are the lr2xx instances [24]. In addition, 60 datasets are generated with the easiest combination, i.e., (a1, c1, s1, tw1, v1), to compare the best GGA variant to Cplex.

5.2 Results

In this study, four variants of the proposed GGA depending on the selection operator and the population management are compared using all generated datasets. The fitness proportional roulette wheel selection (FitPro) and binary tournament (BinT) are combined with the steady-state population management (StSt) and general replacement with elitism (GenRep). In each of the four possible combinations, all other parameters are set as depicted in Table 3 based on Pankratz, since preliminary studies show, that the parameter setting is reasonable. Please note that the variant "BinT_StSt" is equivalent to the GGA introduced in [19]. The GGA is implemented in C++ and executed single threaded on an Ubuntu Server with a 2.1 GHz CPU and 128 GB RAM.

In Fig. 5, the relative error of the objective values for the GGA variants to the best solution among all four variants are calculated for each instance and

Table 2. Used parameters for creating datasets with generator

Number of requests:	(a1) 10, (a2) 15, (a3) 20
Number of clusters:	(c1) 2, (c2) 3, (c3) 4
Mean of cluster radius:	(s1) 5, (s2) 30, (s3) 60
Time window [min, mean, dev]:	(tw1) $[20, 90, 40]$, (tw2) $[150, 600, 400]$
Vehicle capacity:	(v1) $[150, 200, 250]$, (v2) $[300, 400, 500]$, (v3) $[500, 650, 800]$

displayed as boxplots. It can be seen that on average the GGA variants with general replacement outperform the ones with steady-state population management regarding mean and standard deviation (sd) as well as regarding the computation time (cf. Table 4). This implies that GenRep is more often better than StSt and it is less far away from the best fitness value if its solution is not the minimum. A reason for this is that StSt is too intensifying for a sufficiently large search in the solution space without implementing a clever maintenance mechanism and replacement strategy. This is necessary due to the complexity of the considered MDPDPTWHV. In order to find sufficiently good solutions, the diversity of the initial population is much more important in this StSt than in GenRep. Moreover, it can be evaluated that dataset variants 6, 8 and 11 give the worst GGA results. These variants have parameter settings similar to the lrc1xx, lrc2xx and lr2xx Li and Lim datasets, which are also known as the more difficult instances.

Afterwards, the best approach FitPro_GenRep is taken and compared to the best solution found by Cplex for the 60 instances built with the easiest attribute combination (cf. Table 2). To do so, the introduced two-index formulation is implemented in GAMS and run on a Windows Server (2.72 GHz and 383 GB RAM) with 30 Threads in parallel, an execution time limit of 2 h, and an initial solution given by the BinT_StSt variant with $\eta_{max} = 1,000$ individuals. If Cplex's solution is worse than the one of the FitPro_GenRep variant due to running out of the execution time limit, the FitPro_GenRep GGA will be compared to the best solution of all four described GGA variants. While comparing

Table 3. Parameter settings for the comparison of GGA variants

Parameter	Value	Description
n_{pop}	50	Population size
p_{cross}	1.0	Crossover probability
p_{mut}	0.3	Mutation probability
γ_{max}	250	Max. number of generations in GenRep
η_{max}	20,000	Max. number of individuals in StSt
$\tilde{\eta}_{max}$	4,000	Max. number of individuals without improvement in StSt

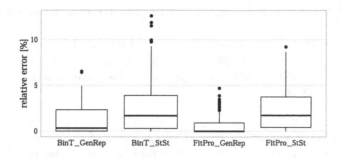

Fig. 5. Boxplot of the relative errors to the best solution variant in %

Table 4. Average relative errors and average computation times of all GGA variants for all instances (single run)

Variant	Fit_error_{mean}	Fit_error_{sd}	$ComputationTime_{mean}$
BinT_GenRep	1.23%	1.57%	74.995 s
BinT_StSt	2.63%	2.86%	85.433 s
FitPro_GenRep	0.62%	1.03%	72.675 s
FitPro_StSt	2.43%	2.35%	87.636 s

the GGA variant FitPro_GenRep to the best known solutions, it is 2.43% worse, whereas FitPro_GenRep is 2.12% worse on average compared to all optimal solutions, as it is presented in Table 5. Furthermore, the computation times of Cplex (in which the instances that have reached the time limit are also integrated) are significantly longer than the computation times of the GGA. Thus, the results show that the developed GGA gives a sufficiently good tool with respect to the solution quality to solve the MDPDPTWHV.

6   Discussion

In this paper, we have considered the Multi Depot Pickup and Delivery Problem with Time Windows and Heterogeneous Vehicle Fleets. To model the problem, a new mathematical two-index formulation has been introduced, which gives the ability to solve more complex instances of the problem with exact solvers, e.g., Cplex, compared to a typical three-index formulation. Due to the high complexity of the MDPDPTWHV, a Grouping Genetic Algorithm has been proposed to solve the addressed problem heuristically, which has not been done in the literature yet. The developed GGA is based on Pankratz [19], who invented a variant for the PDPTW. In addition, the algorithm has been extended so that it can consider heterogeneous vehicles and several depots. In order to evaluate the GGA, self-generated instances were used, which are similar to the well known Li and Lim [18] datasets, and four variants (regarding population management and selection) of the GGA were compared to each other. The best algorithm of these

Table 5. Relative error on average of FitPro_GenRep to best known solution

#exact	Fit_error$_{exact}$	Fit_error$_{all}$	Cplex$_{exe.}$	GGA$_{exe.}$
19	2.12%	2.43%	5330.162 s	72.675 s

Remark. The GGA is evaluated on a less powerful server than Cplex.

has been competed against the best known solution of the datasets (calculated by Cplex or one of the GGA variants). It could be shown, that the steady-state variants performed worse than the general replacement versions due to the high intensification, which has a negative impact on searches in large solution spaces. As a result, the best GGA variant is 2.43% worse than the best known solution.

In future research, the two variants, i.e., general replacement and steady-state, should be combined due to the high intensifying steady-state behavior. In the first stage of the GGA, the general replacement will investigate the solution space in a broad way followed by the second stage, where the population then is intensified by using steady-state for determinating the best individuals. As a further reasonable enhancement of the GGA, an adaptive genetic algorithm approach with respect to the probability values, e.g., for the mutation variants will improve the presented results should be investigated. In addition, it would be interesting to find out whether the algorithm described also produces similarly good results for, e.g., dial-a-ride problems in which people are transported instead of goods.

References

1. Alaia, E.B., et al.: Optimization of the multi-depot & multi-vehicle pickup and delivery problem with time windows using genetic algorithm. In: International Conference on Control, Decision and Information Technologies (CoDIT), Hammamet, Tunisia (2013)
2. Betriebswirtschaft und Operations Research Homepage. https://www.uni-hildes heim.de/fb4/institute/bwl/betriebswirtschaft-und-operations-research/. Accessed 7 Nov 2019
3. Bettinelli, A., Ceselli, A., Righini, G.: A branch-and-price algorithm for the multi-depot heterogeneous-fleet pickup and delivery problem with soft time windows. Math. Program. Comput. **6**(2), 171–197 (2014). https://doi.org/10.1007/s12532-014-0064-0
4. Caracez-Cruz, J., et al.: Rich vehicle routing problem: survey. ACM Comput. Surv. (CSUR) **47**(2), 1–28 (2015)
5. Schiller, T., et al.: Global Truck Study 2016. Deloitte, New York (2017)
6. Desaulniers, G., et al.: VRP with pickup and delivery. In: Toth, P., Vido, D. (eds.) The Vehicle Routing Problem, pp. 225–242. SIAM, Philadelphia (2002)
7. Dondo, R., Cerdà, J.: A cluster-based optimization approach for the multi-depot heterogeneous fleet vehicle routing problem with time windows. Eur. J. Oper. Res. **176**, 1478–1507 (2007)

8. Elbert, R., Thiel, D., Reinhardt, D.: Delivery time windows for road freight carriers and forwarders—influence of delivery time windows on the costs of road transport services. In: Clausen, U., Friedrich, H., Thaller, C., Geiger, C. (eds.) Commercial Transport. LNL, pp. 255–274. Springer, Cham (2016). https://doi.org/10.1007/978-3-319-21266-1_17

9. Eurotransport Homepage. https://www.eurotransport.de/artikel/standgeld-anerk annter-ausgleich-6543390.html. Accessed 26 Oct 2019

10. Falkenauer, E.: Genetic Algorithms and Grouping Problems, 1st edn. Wiley, New York (1998)

11. Frauenhofer SCS: Führende Logistikdienstleister im Bereich Stückgut nach Umsatz in Deutschland im Jahr 2015. Deutsche Verkehrszeitung **82**(9) (2017)

12. Eiben, A.E., Smith, J.E.: Evolutionary robotics. In: Eiben, A.E., Smith, J.E. (eds.) Introduction to Evolutionary Computing. NCS, pp. 245–258. Springer, Heidelberg (2015). https://doi.org/10.1007/978-3-662-44874-8_17

13. Gansterer, M., Hartl, R.F.: Collaborative vehicle routing: a survey. Eur. J. Oper. Res. **268**(1), 1–12 (2018)

14. Goel, A., Gruhn, V.: Large Neighborhood Search for rich VRP with multiple pickup and delivery locations. In: Proceedings of the 18th Mini EURO Conference on VNS (2005)

15. Irnich, S.: A multi-depot pickup and delivery problem with a single hub and heterogeneous vehicles. Eur. J. Oper. Res. **122**, 310–328 (2000)

16. Lahyani, R., et al.: Rich vehicle routing problems: from a taxonomy to a definition. Eur. J. Oper. Res. **241**, 1–14 (2015)

17. Li, Y., et al.: Adaptive large neighborhood search for the pickup and delivery problem with time windows, profits, and reserved requests. Eur. J. Oper. Res. **252**, 27–38 (2016)

18. Li, H., Lim, A.: A metaheuristic for the pickup and delivery problem with time windows. Int. J. Artif. Intell. Tools **12**(2), 160–167 (2001)

19. Pankratz, G.: A grouping genetic algorithm for the pickup and delivery problem with time windows. OR Spectr. **27**(1), 21–41 (2005). https://doi.org/10.1007/s00291-004-0173-7

20. Razali, N.M., Geraghty, J.: Genetic algorithm performance with different selection strategies in solving TSP. In: 2011 Proceedings of the World Congress on Engineering, vol. II, pp. 1134–1139. Newswood Limited, London (2011)

21. Ropke, S., Pisinger, D.: An adaptive large neighborhood search heuristic for the pickup and delivery problem with time windows. Transp. Sci. **40**(4), 455–472 (2006)

22. Savelsbergh, M.W.P., Sol, M.: The general pickup and delivery problem. Transp. Sci. **29**(1), 17–29 (1995)

23. Schwemmer, M.: Top 100 der Logistik - Marktgrößen, Marktsegment, Marktführer. Frauenhofer IIS (2016)

24. SINTEF. https://www.sintef.no/projectweb/top/pdptw/li-lim-benchmark/. Accessed 14 Nov 2019

25. Sombuntham, P., Kachitvichyanukul, V.: Multi-depot vehicle routing problem with pickup and delivery requests. AIP Conf. Proc. **1285**, 71 (2010)

26. Syswerda, G.: A study of reproduction in generational and steady-state genetic algorithms. Found. Genet. Algorithms **1**, 94–101 (1991)

27. Wen, M.: Rich vehicle routing problems and applications. Ph.D. thesis, DTU Management Engineering (2010)

MILPIBEA: Algorithm for Multi-objective Features Selection in (Evolving) Software Product Lines

Takfarinas Saber[1(✉)], David Brevet[2], Goetz Botterweck[3], and Anthony Ventresque[1]

[1] School of Computer Science, University College Dublin, Dublin, Ireland
{takfarinas.saber,anthony.ventresque}@ucd.ie
[2] Laboratoire d'Informatique, de Modélisation et d'Optimisation des Systèmes, Aubiere, France
david.brevet@uca.fr
[3] University of Limerick, Limerick, Ireland
goetz.botterweck@lero.ie

Abstract. *Software Product Lines Engineering* (SPLE) proposes techniques to model, create and improve groups of related software systems in a systematic way, with different alternatives formally expressed, e.g., as *Feature Models*. Selecting the 'best' software system(s) turns into a problem of improving the quality of selected subsets of software features (components) from feature models, or as it is widely known, *Feature Configuration*. When there are different independent dimensions to assess how good a software product is, the problem becomes even more challenging – it is then a multi-objective optimisation problem. Another big issue for software systems is evolution where software components change. This is common in the industry but, as far as we know, there is no algorithm designed to the particular case of multi-objective optimisation of evolving software product lines. In this paper we present MILPIBEA, a novel hybrid algorithm which combines the scalability of a genetic algorithm (IBEA) with the accuracy of a mixed-integer linear programming solver (IBM ILOG CPLEX). We also study the behaviour of our solution (MILPIBEA) in contrast with SATIBEA, a state-of-the-art algorithm in static software product lines. We demonstrate that MILPIBEA outperforms SATIBEA on average, especially for the most challenging problem instances, and that MILPIBEA is the one that continues to improve the quality of the solutions when SATIBEA stagnates (in the evolving context).

Keywords: Software product line · Feature selection · Multi-objective optimisation · Evolutionary algorithm · Mixed-integer linear programming

L. Paquete and C. Zarges (Eds.): EvoCOP 2020, LNCS 12102, pp. 164–179, 2020.
https://doi.org/10.1007/978-3-030-43680-3_11

1   Introduction

Software Engineering combines various domains [1]. Software Product Lines (SPL) is one of these domains that deal with groups of related software systems as an ensemble, instead of handling each of them independently [2]. SPL is getting more attention by the software industry as it simplifies software reuse [3] and enables better reliability and important reduction in cost [4].

A common way to represent a product line, all available products and their essential characteristics, is a Feature Model (FM). Every feature corresponds to an element of a software system/product that is of interest to a particular company. Each FM describes the available configuration choices, and consequently the set of all possible products as combinations of features. These FMs can grow to become very large (e.g., in this paper, we use FMs with ∼7k features and ∼350k constraints).

When deriving a particular product from the product line, we have to perform a feature selection. To find the best possible product, we *optimise the feature selection*, i.e., pick the set of features which gives us the 'best' product [5]. Since in practice various characteristics often have to be considered simultaneously (e.g., cost, technical feasibility, or reliability) finding the 'best' feature selection is an instance of a *multi-objective optimisation problem* [6].

A similar problem that is not fully studied in the literature is *the multi-objective feature selection* when FMs evolve. There is a continual evolution of software libraries and a constant change in customers' preferences regarding the requirements of software applications. These evolutions appear as an adaptation of the FM from a version to another. For instance, Saber et al. [7] have shown in their study that the large FM representing the Linux kernel evolves continuously. They have also shown that a new version of the kernel is released every few months with a successive difference that can go up to 7%.

In this paper, we propose to leverage the evolution context when performing optimisations of feature configurations. It seems odd to generate random bootstrapping populations for SATIBEA in the presence of well-performing solutions for similar problem instances. At the same time, it might be beneficial to exploit the fact that the FM has evolved and that configurations generated previously are close enough and can be adapted.

This paper presents our approach, MILPIBEA, which was initially designed to address the problem of feature selection in a multi-objective context when the FMs evolve, but proved to be better than SATIBEA both when the FMs evolve and when they do not. MILPIBEA is a hybrid algorithm that uses a genetic algorithm (IBEA) and a mixed-integer linear programming (MILP) solver (IBM ILOG CPLEX).

SATIBEA [6] (also a hybrid algorithm) faces a difficult challenge: the search space is so large and constrained that mutation and crossover operations generate a large number of infeasible solutions. SATIBEA uses a SAT solver to fix these infeasible solutions and obtains (close to) viable individuals at each generation of the genetic algorithm. However, this process has two major issues: (i) it is time-consuming – an empirical study of SATIBEA showed that the vast majority of the execution time consists in fixing the faulty individuals; (ii) it modifies the individuals, often substantially, which defies the idea behind genetic algorithms – where you expect to inherit properties from previous generations and modify the individuals only marginally. MILPIBEA's correction of individuals is both more efficient and more effective, making sure the corrected individuals are closer to the ones generated by IBEA's mutation and crossover.

This paper makes the following contributions:

- We propose MILPIBEA, a hybrid algorithm that outperforms the SAT-IBEA [6] both in terms of execution time and quality of the solutions.
- We thoroughly evaluate SATIBEA and MILPIBEA on evolving and non-evolving SPL problems and show that MILPIBEA is 42% better than SAT-IBEA in hypervolume on average, especially for the most challenging problem instances, and that MILPIBEA is the one that continues to improve the quality of solutions when SATIBEA stagnates (in the evolving context) and does improve the quality of solutions.

Combining a solver with a multi-objective evolutionary algorithm has already been proposed to address the particular problem of multi-objective feature selection in SPL [6,8,9] and problems from various other problem domains (e.g., cloud computing [10–12]). However, this is the first work that proposes using a MILP solver for the multi-objective feature selection in SPL.

The remainder of this paper is organised as follows: Sect. 2 describes the context of our study. Section 3 provides the overall set-up and the benchmark for evolving SPL. We then discuss potential improvements in three steps: Sect. 4 motivates seeding previous solutions when dealing with evolving FMs. Section 5 compares the correction mechanisms of SATIBEA and MILPIBEA. Section 6 discusses how MILPIBEA performs in comparison to SATIBEA in terms of achieved hypervolume and required time. Finally, Sect. 7 concludes the paper.

2   Background

In this section, we present four elements that form our research's background:

- Software Product Line Engineering, in particular how to describe variations of software applications as configurations of a feature model.
- Multi-objective optimisation (MOO); picking features can lead to many products for which the quality can be seen from different perspectives. MOO gives a framework to address this sort of problems.
- Evolution of Software Product Lines: Software applications, requirements, and implementations change constantly. Therefore, feature models need to be updated to reflect these evolutions [7].
- SATIBEA, a state-of-the-art algorithm to address the MOO for feature selection in feature models [6] and the same when FMs evolve [7].

2.1 Software Product Line Engineering

Software engineers often need to adapt software artefacts to the needs of a particular customer. Software Product Line Engineering (SPLE) is a software paradigm that aims at managing those variations in a systematic fashion. For instance, all software artefacts (and their variations) can be interpreted as a set of features which can be selected and combined to obtain a particular product.

Feature Models can be represented as a set of features and connecting relationships (constraints). Figure 1 shows a toy FM which has ten features connected by several relationships. For instance, each 'Screen' has to be have exactly one of three types, i.e., 'Basic', 'Colour' or 'High Resolution'. When deriving a software product from the software product line, we have to select a subset of features $\mathcal{S} \subseteq \mathcal{F}$ that satisfies the FM $\mathcal{F}$ and the requirements of the stakeholder/customer. This configuration can be described as a satisfiability problem (SAT), i.e., instantiating variables (in our case, features) with the values true or false in a way that satisfies all the constraints. Let $f_i \in \{$true, false$\}$ which is set to true if the feature $F_i \in \mathcal{F}$ is selected to be part of $\mathcal{S}$ and false otherwise.

An FM is represented in a conjunctive normal form (CNF). Finding a product in the SPL is then equivalent to assigning a value in $\{$true, false$\}$ to every feature. For instance, in Fig. 1 the FM would have the following clauses, among others: $(Basic \vee Colour \vee High\ resolution) \wedge (\neg Basic \vee \neg Colour) \wedge (\neg Basic \vee \neg High\ resolution) \wedge (\neg Colour \vee \neg High\ resolution)$, which describe the alternative between the three screen features. When configuring a SPL, software designers do not limit themselves to finding possible products (satisfying the FM) but also attempt to discover products optimising multiple criteria. For this reason, SPL configuration is modelled as multi-objective problem.

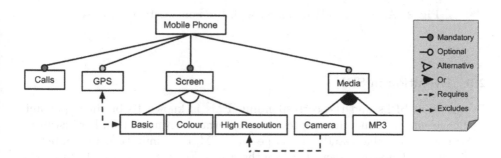

Fig. 1. Example of a feature model

2.2 Multi-objective Optimisation

Multi-Objective Optimisation (MOO) involves the simultaneous optimisation of more than one objective function. Given that the value of software artefacts can be seen from different angles (e.g., cost, importance, reliability), feature selection in SPL is a good candidate for MOO [6].

Solutions of a MOO problem represent the set of non-dominated solutions defined as follows: Let S be the set of all feasible solutions for a given FM. Then $\forall x \in S$, $F = [O_1(x), ..., O_k(x)]$ represents a vector containing values of the k objective functions for a given solution x. We say that a solution x_1 dominates x_2, written as $x_1 \succ x_2$, if and only if $\forall i \in \{1, ..., k\}$, $O_i(x_1) \leq O_i(x_2)$ and $\exists i \in \{1, ..., k\}$ such that $O_i(x_1) < O_i(x_2)$. We also say that x_i is a non-dominated solution if there is no other solution x_j in the Pareto front s.t. x_j dominates x_i.

All the non-dominated solutions represent a set called a Pareto front: in this set, it is impossible to find any solution better in all objectives than the other solutions in the set. The Pareto front given in Fig. 2 contains solutions x_1, x_2, x_4, x_6, x_7 because they are not dominated by any other, while, for instance, x_8 is dominated by x_1. Hence, x_8 is not in the Pareto front.

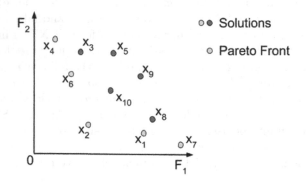

Fig. 2. Example of a Pareto front with two minimisation objectives.

2.3   Evolution in SPL

Evolution of SPLs and the corresponding FMs is known to be an important challenge, since product lines represent long-term investments [13]. For instance, in Sect. 3 we describe a study of a large-scale FM, the Linux kernel by Saber et al. [7] which shows that every few months a new FM is released with up to 7% modifications among the features (features added or removed).

In this paper, we show a potential approach for this optimisation problem which utilises the evolution from one FM to another. The relationship between two versions of a feature model is expressed as a mapping between features. Let us assume an FM FM_1 evolved into another FM FM_2. Some of the features $f_i^1 \in FM_1$ are mapped on to features $f_i^2 \in FM_2$ (treated as the same), whereas other features $f_i^1 \in FM_1$ are not mapped onto any features in FM_2 (f_i^1 has been removed), and features $f_i^2 \in FM_2$ have no corresponding features in FM_1 (f_i^2 has been added). The same can be applied to constraints (removed from FM_1 or added to FM_2). The problem we address concerns adapting the solutions found previously for FM_1 to FM_2.

2.4 SATIBEA

SATIBEA [6] is an extension of the Indicator-Based Evolutionary Algorithm (IBEA) which guides the search by a quality indicator given by the user. Previously to SATIBEA, several techniques have been tried to solve the multi-objective feature selection in SPL. As most of the random techniques and genetic algorithms tend to generate invalid solutions (given the large and constrained search space, any random, mutation or crossover operation is tricky) setting the number of violated constraints as a minimisation objective has been proposed by Sayyad et al. [14] and has since been widely used in the literature [6–8]. It is not the best possible decision and is acceptable only for small problems (only small FMs are solvable with exact algorithms [15]).

SATIBEA has been introduced to help IBEA find valid products using a SAT solver. SATIBEA changes the mutation process of IBEA: when an individual is mutated, three different exclusive mutations can be applied:

1. The standard bit-flip mutation proposed by IBEA.
2. Replacing the individual by another one generated by the SAT solver that does not violate any constraints.
3. Transforming the individual into a valid one using the SAT solver (repair).

Using this novel mutation approach, SATIBEA finds better solutions than IBEA: it finds valid optimised products, but also gives better values in quality metrics.

In this paper, we propose MILPIBEA; a novel technique that addresses some of SATIBEA's limitations (i.e., slow and stagnating performance improvements).

3 System Set-Up

This section presents the different elements that we have used in our implementation: the data set, the objectives we use for our multi-objective optimisation problem, the metric we use (i.e., hypervolume), the parameters we use for the genetic algorithm (i.e., IBEA) and the hardware configuration.

3.1 Benchmark for Evolving FMs

Our work is based on the largest open-source FM we could find in the literature: the Linux kernel version 2.6.28 containing 6,888 features and 343,944 constraints.

Saber et al. [7,16] studied the demographics (features/constraints) and evolution pattern of 21 successive versions of the Linux kernel (going from 2.6.12 to 2.6.32). They observed that on average there was only 4.6% difference in terms of features between a version and the next (out of those changes, 21.22% were removed features and 78.78% were added features). They also evaluated the size of the clauses/constraints in the problem, as we need to know how the constraints we add in the problem should look and found that a large proportion of the FMs' constraints have 6 features (39%), 5 features (16%), 18 features (14%) or 19 features (14%). Saber et al. [7] put at our disposal a generator of synthetic FM evolutions based on the real evolution of the Linux kernel – hence a realistic benchmark but with more variability than in a real one, allowing us also to get several synthetic data sets corresponding to these characteristics.

The FM generator provided by Saber et al. uses two parameters representing the percentage of feature modifications (added/removed) and the percentage of constraint modifications (added/removed). The higher those percentages are, the more different the new FM is from its original. The FM generator uses the proportions observed in the 20 FMs to generate new features/remove old ones, and to generate new constraints of a particular length. We use the following values to generate evolved FMs: from 5% of modified features and 1% of modified constraints (FM 5_1) to 20% of modified features and 10% of modified constraints (FM 20_10). In our evaluations we generate 10 synthetic FMs for each parameter values. Data is available at https://github.com/aventresque/EvolvingFMs.

3.2   Optimisation Objectives

We use a set of optimisation objectives from the literature [6]:

1. *Correctness* – minimise the number of violated constraints, proposed by Sayyad et al. [14].
2. *Richness of features* – maximise the number of selected features (have products with more functionality).
3. *Features used before* – minimise the number of selected features that were not used before.
4. *Known defects* – minimise the number of known defects in selected features (we use random integer values between 0 and 10).
5. *Cost* – minimise the cost of the selected features (we use random real values between 5.0 and 15.0).

In a different application context, these objectives could be augmented or replaced with other criteria, e.g., consumption of resources or various costs.

3.3   Hypervolume Indicator

We evaluate the quality of our solutions cost using the hypervolume metric [17]. Intuition behind the hypervolume is that it gives the volume (in the k dimensions of the search space) dominated by a set of non-dominated solutions. Hypervolume is the region between the solutions and the reference point (the higher the better). The reference point is set with the worst value for each of the objectives.

3.4 System and Algorithms Set-Up

We use the source code provided by SATIBEA's authors and make MILPIBEA publicly available at https://github.com/takfarinassaber/MILPIBEA. The tests are performed on a machine with 62 GB of RAM and 12 core Intel(R) Xeon(R) 2.20 GHz CPU. We use the following parameters for our genetic algorithm:

- Population size: 300 individuals.
- Offspring population size: 300 individuals.
- Crossover rate: 0.8. Represents the probability of two individuals in the population to perform a crossover (an exchange of their selected features).
- Mutation rate: 0.001. Represents the probability for each bit (true if a feature is selected, 0 otherwise) of an individual to be flipped.
- Solver mutation rate: 0.02. Represents the probability of using the solver to correct a solution during the mutation process.

We also use one heuristic in our algorithm: we do not do any bit flip for mandatory or dead features as this always lead to invalid products. We use the engine of the MILP solver *IBM ILOG CPLEX*. We use the hypervolume metric proposed by Fonseca et al. [17]. We ran all our algorithm instances for 20 min and determined the average over 10 runs (for each randomly generated instance).

4 Using Seeds in Evolving FM

In this section, we explore how to use seeds (including previously found solutions in the initial population of a new evolution) to take advantage of the fact that the feature model evolved.

When a FM evolves, the modifications of features and constraints depend on how different the two models are (new and original models). We propose to take advantage of previous FM configurations (when they exist) to feed SATIBEA with solutions of the original model. Let's suppose two FMs: F_1 and F_2 with F_2 being an evolution of F_1 (i.e., features/constraints added and removed). We consider that we already found a set of solutions S_1 by applying a multi-objective optimisation algorithm (SATIBEA in our case) on F_1. Instead of leaving SATIBEA with an initial random population for F_2 (similar to what is proposed in [7]), we adapt S_1 to F_2. Therefore, for each individual, we remove bits representing removed features and add bits with random values for each new feature. Then, we compute their objective functions. We give the new resulting individuals as an initial population to SATIBEA that will run normally on F_2. Our hope is that initial individuals will be better than random solutions.

We tested this approach on all the modified versions of the Linux Kernel and all of the results are equivalent: as expected, when supplied with an initial seed SATIBEA converges within a short time (i.e., less than 100 s) whereas the classical SATIBEA needs 700 s to reach the same hypervolume. This approach also has some limits: with a modified version of 20% features and 10% constraints, classical SATIBEA reaches a slightly better hypervolume than the one with seed.

When we give an initial population to SATIBEA, it converges very fast. Still, it is also blocked very fast, i.e., after 100 s on all models, it stagnates and is unable to improve results further. That is why we decided to investigate a better substitution for SATIBEA, starting from its repair technique based on a SAT solver. We describe this approach in the next section.

5 Correcting Individuals

In this section, we present two ways of correcting non-feasible individuals, i.e., a situation that happens very often during the execution of the genetic algorithm for our problem. Indeed, both mutation and crossover, the basic operations of (SAT)IBEA, generate quite a large ratio of infeasible individuals, given the size of the search space and the number of constraints that can be violated.

The first solution we present is the one proposed in the definition of SAT-IBEA [6]. The second one is our own improved solution using the MILP solver. Eventually, we propose an evaluation of the two techniques with an example.

5.1 How SATIBEA Corrects Solutions

SATIBEA's correction method occurs in the mutation phase of the genetic algorithm. IBEA takes an individual that violates one or several constraints out of the population and corrects it, using a SAT solver. This leads to the individual being now valid (no longer violating constraints). Figure 3 shows an example of SATIBEA's repair technique on a FM with 5 features (f_1 to f_5) and 3 constraints (c_1 to c_3). The constraints are shown on the left-hand side of Fig. 3, with c_2 marked as violated.

(1) First, an individual with assignment {$1\ 1\ 1\ 0\ 0$} is selected for repair due to the violation of constraints c_2 (which causes the individual to be invalid). This is shown in row (1) in the table on the right-hand side of Fig. 3.

(2a) Second, SATIBEA unsets (this is represented by '_' in the example) all the bits that belong to a violated constraint. Here, constraint c_2 is violated, so f_4 and f_5 are unset. This is in row ($2a$) of the table.

(2b) Third, SATIBEA unsets all the bits that are evaluated as 'false' in every constraint. Each of these can either be a feature without a negation sign in the constraint (i.e., f) that is set to false or a feature with a negation (i.e., $\overline{f}$) that is set to true. All of these are unset. In our example, f_2 is assigned to true and is evaluated at false in the constraint c_1 ($\overline{f}_2$). Therefore, SATIBEA unsets f_2. This is shown in row ($2b$) of the table.

(3) Eventually, the resulting partial assignment is given to the SAT solver to complete the unset values while satisfying the constraints of the FM. SATIBEA's correction always obtains a valid solution if it exists. In our case, SATIBEA results in a new individual (i.e., {$1\ 0\ 1\ 1\ 0$}). This is shown on line 3 of Fig. 3. Note that this procedure cannot guarantee to always return a valid individual as the problem may be unsatisfiable.

		f_1	f_2	f_3	f_4	f_5
c_1 $\left(f_1 \vee \bar{f}_2 \vee f_3\right) \wedge$	1)	1	1	1	0	0
$c_{2 \text{ (Violated)}}$ $\left(f_4 \vee f_5\right) \wedge$	2a)	1	1	1	_	_
c_3 $\left(\bar{f}_2 \vee f_3 \vee \bar{f}_5\right)$	2b)	1	_	1	_	_
	3)	1	0	1	1	0

Fig. 3. Correction of an individual in SATIBEA. The original individual, violating constraint 2, is shown on line 1 and the different steps of SATIBEA's correction are shown on lines 2a, 2b and 3.

Although this correction technique is fast and improves the classical IBEA algorithm, the number of flipped bits is large. This often creates new individuals that are far from the original ones (before the correction). This issue is that those individuals were obtained by mutation in IBEA and modifying them too much is against the idea behind genetic algorithms (i.e., inheriting and preserving good characters). For instance, from the individual {1 1 1 0 0} (line 1 of Fig. 3) that violates the constraints, it would be better to obtain individual {1 1 1 1 0} that does not violate the constraints (instead of {1 0 1 1 0}). The next subsection describes our MILP-based correction technique that overcomes this problem.

5.2 How Our MILP Technique Corrects Solutions

Our new method corrects individuals and avoids the problem described in previous section (i.e., a large number of flipped bits between the initial individuals and the corrected ones). This method corrects the faulty individuals and minimises the number of flipped bits which are not part of any violated constraint.

Applied to the example in Fig. 3, only features f_4 and f_5 are unset. CPLEX solves the problem of finding a valid individual by assigning values to f_4 and f_5 while at the same time, minimising the total bit flips on the rest of the features (i.e., f_1, f_2 and f_3). One possible output is {1 1 1 1 0} which does not modify any fixed bit, unlike SATIBEA's one (i.e., {1 0 1 1 0} which has one modification on the feature f_2).

Using our method, CPLEX is guaranteed to find a valid individual. Moreover, it returns an individual that is as close as possible to the original one. In our method, we use the model defined by Eqs. 1a, 1b, and 1c.

$$\textit{Minimise} \qquad \sum_{x \in T}(1 - x) + \sum_{x \in F} x \qquad (1a)$$

$$\text{Subject to} \qquad \sum_{x \in P_i} x + \sum_{x \in N_i}(1 - x) \geq 1, \quad \forall i \in \{1, .. n\} \qquad (1b)$$

$$x \in \{0, 1\}, \quad \forall x \in X \qquad (1c)$$

With n number of clauses, X set of features, $T \subset X$ set a features fixed at true, $F \subset X$ set of features fixed at false, $P_i \subset X$ set of features without negation in clause i, and $N_i \subset X$ set of features with negation in clause i.

In the MILP model above, we aim to minimise the number of flipped features that were not part of violated constraints in the original individual: if the feature was originally at True (i.e., '1'), then we count it as a modification if and only if it changes to False (i.e., '0'). Similarly, when the feature was originally at False and is changed to True, we also count it as a modification. As in the technique using the SAT model, each clause is represented by a linear constraint. Every feature without a negation is considered as '1' when selected, and every feature that is negated is considered as '1' when unselected. The sum of every feature within a clause has to be larger or equal to 1 to validate it.

Table 1. Comparison of SATIBEA and MILP corrections. Higher values of hypervolume (HV) are better. Lower values of both time and number of modifications (#mod) are better. Best values for each instance in bold.

Instance	No correction	SATIBEA correction			MILP correction		
	HV	HV	Time (ms)	#mod	HV	Time (ms)	#mod
1_1	1.09	1.12	8,895	2,696	**1.18**	**6,801**	**141**
5_1	2.15	2.27	8,474	2,660	**2.35**	**2,337**	**353**
5_3	1.03	1.14	9,005	2,747	**1.25**	**5,675**	**297**
10_1	1.00	1.04	9,273	2,615	**1.16**	**2,192**	**798**
10_3	1.08	1.19	10,255	2,732	**1.33**	**4,044**	**99**
10_5	1.02	1.18	10,339	2,784	**1.33**	**4,082**	**110**
20_1	0.90	0.96	9,762	2,528	**1.03**	**2,143**	**412**
20_3	2.16	2.28	9,699	2,691	**2.47**	**2,891**	**148**
20_5	1.06	1.20	10,222	2,752	**1.36**	**2,962**	**124**
20_10	0.72	0.73	10,877	3,008	**0.75**	**6,719**	**92**

5.3   Comparison with Respect to the Correction Process

In Table 1, we compare our correction method against SATIBEA's correction. Each instance corresponds to an evolved FM and is represented by a couple (x_y) where x is the percentage of features modified and y the percentage of constraints modified. We took the 300 individuals given by SATIBEA on the original FM as seeds for the evolved versions. SATIBEA found 62 solutions that do not violate constraints in the original FM. Obviously, these solutions violate some constraints in each of the evolved FMs. We compared both SATIBEA's and MILP's correction methods applied on the 62 individuals. We measured the hypervolume (HV), the average execution time for each individual (in milliseconds) and the average number of modified features from the original individual to the corrected one (#mod). We also added the hypervolume of non-corrected solutions (NC) as a baseline.

We can see that when applying a correction, both algorithms improve the hypervolume composed to non-corrected individuals. However, MILP's corrections outperforms SATIBEA's corrections. Correction using a MILP solver improved hypervolume of SATIBEA's correction by 147% on average, while only requiring 48% of its execution time. Moreover, we notice that the number of modified features per individual using the MILP correction is one order of magnitude lower than when using SATIBEA's correction (on average SATIBEA's correction requires 2,721 feature modifications whereas MILP's correction only requires 257). As our correction method needs less modifications of individuals to transform them into valid ones, it could be more interesting to use it instead of SATIBEA's one in a genetic algorithm: indeed, less modifications imply a better conservation of the accumulated knowledge during the generations. An implementation of our genetic algorithm with this type of correction method in the mutation part is described in the next section.

6   Performance of MILPIBEA vs. SATIBEA

We now report on how MILPIBEA and SATIBEA perform on the multi-objective features selection problem – in particular with respect to the achieved hypervolume and the required time for that. We initially discuss the general feature selection problem and then the case of evolved FMs.

6.1   On the Multi-objective Feature Selection Problem

Figure 4 show the evolution using SATIBEA or MILPIBEA in terms of hypervolume when applied on our 10 generated models: each of them is a modification of the 2.6.28 version of Linux kernel represented by a couple (x_y) where x is the percentage of features modified and y the percentage of constraints modified. This hypervolume is measured based only on the individuals of the current population. We are not seeding the initial population: the problem studied in these result is the multi-objective feature selection problem, without the notion of evolution. The initial population is generated randomly for both algorithms.

Our results indicate that MILPIBEA outperforms SATIBEA with an improvement of 41.2% hypervolume on average. Figure 4 also indicates that MILPIBEA is more efficient on the most constrained problems (i.e., with constraint modifications $\geq 5\%$). MILPIBEA reaches a good hypervolume after 100 s, then increases slowly. We can see that SATIBEA's hypervolume increases with a slower pace than MILPIBEA's; then its hypervolume stays stable (within a small interval).

6.2   With Evolved Feature Models

We now compare MILPIBEA and SATIBEA in the case of the multi-objective feature selection problem in evolving FMs. As described in Sect. 2, the notion of

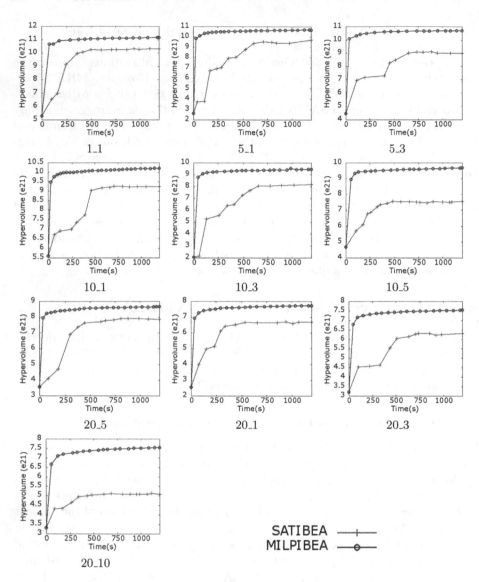

Fig. 4. Comparison of MILPIBEA and SATIBEA on various evolved FMs. The higher the better for the hypervolume.

evolution is represented by features/constraints modifications in the FM. In our case, the Linux kernel 2.6.28 is the original FM, and we generated 10 modified versions. Because of evolution, the original FM has been optimised, and its solutions as given as initial population to SATIBEA and MILPIBEA: the purpose is to improve the quality of results on modified FMs as fast as possible.

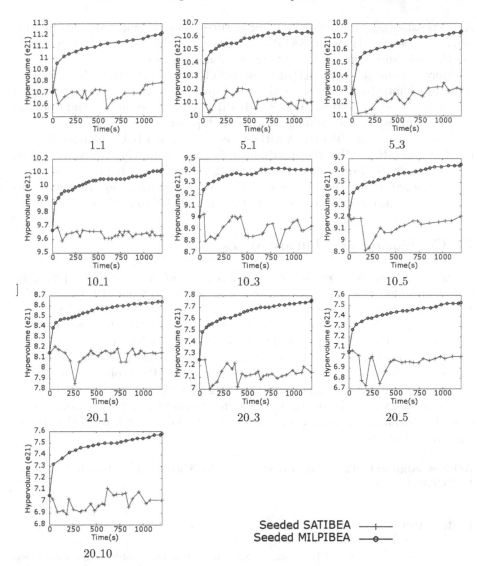

Fig. 5. Comparison of the hypervolume achieved by seeded MILPIBEA and seeded SATIBEA on various evolved FMs (higher values are better).

Figure 5 show the hypervolume of the individuals at every new generation for both SATIBEA and MILPIBEA when given the solutions of the original FM as initial population. We see that both algorithms start from a relatively good hypervolume, which shows the quality of the initial population.

We also see that MILPIBEA successfully improves the hypervolume, whereas SATIBEA struggles when seeded. This is mainly because MILPIBEA has a correction method that allows it to take advantage of the initial population's good characteristics by not changing a lot of features in individuals that are obtained from the crossover. However, SATIBEA requires to modify several features, making the individuals obtained by the repair almost random.

Moreover, we can observe that unlike SATIBEA, MILPIBEA's hypervolume continues improving slowly even after the limit (i.e., 1200 s). A larger allowed time would lead to better solutions. MILPIBEA stagnates after 40 min beyond which we might consider adding a local search phase [18–21].

When comparing MILPIBEA without seeds and MILPIBEA with seeds: after the first generation, MILPIBEA with seeds is 10.5% better in hypervolume than without seeds. It also reaches 97.28% of MILPIBEA's final hypervolume (computed in 1200 s) after only one generation (42.29 s on average). This shows us that a good initial population improves the time needed to reach good solutions.

7   Conclusion and Future Work

In this paper, we have presented the importance of the evolution in SPL by introducing the multi-objective features selection in evolving SPL problem. To solve this problem, we proposed a method based on a combination of a genetic algorithm (IBEA) with a MILP solver (i.e., CPLEX). We observed that this method not only outperforms SATIBEA on the multi-objective features selection but also achieves faster better results in the context of evolving SPL. Our thorough evaluation shows the importance of using a MILP solver to reduce the number of modifications when correcting an individual.

Our future work will investigate the performance with respect to other multi-objective performance metrics and the utility of a local search when the genetic algorithm stagnates.

Acknowledgment. This work was supported by Science Foundation Ireland grant 13/RC/2094.

References

1. Ramirez, A., Romero, J.R., Ventura, S.: A survey of many-objective optimisation in search-based software engineering. J. Syst. Softw. **149**, 382–395 (2019)
2. Metzger, A., Pohl, K.: Software product line engineering and variability management: achievements and challenges. In: FSE, pp. 70–84 (2014)
3. Neto, J.C., da Silva, C.H., Colanzi, T.E., Amaral, A.M.M.M.: Are mas profitable to search-based PLA design? IET Softw. **13**(6), 587–599 (2019)
4. Nair, V., et al.: Data-driven search-based software engineering. In: MSR, pp. 341–352 (2018)
5. Harman, M., Jia, Y., Krinke, J., Langdon, W.B., Petke, J., Zhang, Y.: Search based software engineering for software product line engineering: a survey and directions for future work. In: SPLC, pp. 5–18 (2014)
6. Henard, C., Papadakis, M., Harman, M., Le Traon, Y.: Combining multi-objective search and constraint solving for configuring large software product lines. In: ICSE, pp. 517–528 (2015)

7. Saber, T., Brevet, D., Botterweck, G., Ventresque, A.: Is seeding a good strategy in multi-objective feature selection when feature models evolve? Inf. Softw. Technol. **95**, 266–280 (2018)
8. Guo, J., et al.: Smtibea: a hybrid multi-objective optimization algorithm for configuring large constrained software product lines. Softw. Syst. Model. **18**(2), 1447–1466 (2019)
9. Yu, H., Shi, K., Guo, J., Fan, G., Yang, X., Chen, L.: Combining constraint solving with different MOEAs for configuring large software product lines: a case study. In: COMPSAC, vol. 1, pp. 54–63 (2018)
10. Saber, T., Marques-Silva, J., Thorburn, J., Ventresque, A.: Exact and hybrid solutions for the multi-objective VM reassignment problem. IJAIT **26**(01), 1760004 (2017)
11. Saber, T., Ventresque, A., Marques-Silva, J., Thorburn, J., Murphy, L.: Milp for the multi-objective VM reassignment problem. ICTA **I**, 41–48 (2015)
12. Saber, T., Gandibleux, X., O'Neill, M., Murphy, L., Ventresque, A.: A comparative study of multi-objective machine reassignment algorithms for data centres. J. Heuristics **26**(1), 119–150 (2019). https://doi.org/10.1007/s10732-019-09427-8
13. Pleuss, A., Botterweck, G., Dhungana, D., Polzer, A., Kowalewski, S.: Model-driven support for product line evolution on feature level. J. Syst. Softw. **85**(10), 2261–2274 (2012)
14. Sayyad, A.S., Menzies, T., Ammar, H.: On the value of user preferences in search-based software engineering: a case study in software product lines. In: ICSE, pp. 492–501 (2013)
15. Xue, Y., Li, Y.F.: Multi-objective integer programming approaches for solving optimal feature selection problem: a new perspective on multi-objective optimization problems in SBSE. In: ICSE, pp. 1231–1242 (2018)
16. Brevet, D., Saber, T., Botterweck, G., Ventresque, A.: Preliminary study of multi-objective features selection for evolving software product lines. In: Sarro, F., Deb, K. (eds.) SSBSE 2016. LNCS, vol. 9962, pp. 274–280. Springer, Cham (2016). https://doi.org/10.1007/978-3-319-47106-8_23
17. Fonseca, C.M., Paquete, L., López-Ibáñez, M.: An improved dimension-sweep algorithm for the hypervolume indicator. In: CEC, pp. 1157–1163 (2006)
18. Shi, K., et al.: Mutation with local searching and elite inheritance mechanism in multi-objective optimization algorithm: a case study in software product line. Int. J. Softw. Eng. Knowl. Eng. **29**(09), 1347–1378 (2019)
19. Saber, T., Delavernhe, F., Papadakis, M., O'Neill, M., Ventresque, A.: A hybrid algorithm for multi-objective test case selection. In: CEC, pp. 1–8 (2018)
20. Saber, T., Ventresque, A., Brandic, I., Thorburn, J., Murphy, L.: Towards a multi-objective VM reassignment for large decentralised data centres. In: UCC, pp. 65–74 (2015)
21. Saber, T., Ventresque, A., Gandibleux, X., Murphy, L.: GenNePi: a multi-objective machine reassignment algorithm for data centres. In: HM, pp. 115–129 (2014)

A Group Genetic Algorithm for Resource Allocation in Container-Based Clouds

Boxiong Tan$^{(\boxtimes)}$, Hui Ma, and Yi Mei

Victoria University of Wellington, Wellington, New Zealand
{Boxiong.Tan,Hui.Ma,Yi.Mei}@ecs.vuw.ac.nz

Abstract. Containers have gain popularity because they support fast development and deployment of cloud-native software such as microservices and server-less applications. Additionally, containers have low overhead, hence they save resources in cloud data centers. However, the difficulty of the *Resource Allocation in Container-based clouds (RAC)* is far beyond Virtual Machine (VM)-based clouds. The allocation task selects heterogeneous VMs to host containers and consolidate VMs to Physical Machines (PMs) simultaneously. Due to the high complexity, existing approaches use simple rule-based heuristics and meta-heuristics to solve the *RAC* problem. They either prone to stuck at local optima or have inherent defects in their indirect representations. To address these issues, we propose a novel *group genetic algorithm (GGA)* with a direct representation and problem-specific operators. This design has shown significantly better performance than the state-of-the-art algorithms in a wide range of test datasets.

Keywords: Cloud resource allocation · Container placement · Energy consumption · Group genetic algorithm

1 Introduction

Container-based clouds [14] have quickly become a new trend in cloud computing. Compared to Virtual Machines (VMs), containers (e.g. docker) cause much fewer overheads. This feature is critical for modern cloud-native applications, such as microservices and serverless applications, as they are developed in a decoupling and scalable manner. Cloud providers apply server consolidation [21] strategies in resource allocation to improve the utilization of cloud resources. Server consolidation strategies aim to allocate applications to a minimum number of Physical Machines (PMs), to reduce energy consumption. In container-based clouds, it is much difficult than in VM-based clouds because of the higher granularity of the allocation problem. Server consolidation in VM-based clouds involves one level of allocation, i.e. a set of VMs is allocated to PMs directly while container-based clouds involve two levels of allocation, i.e. a set of containers is allocated to a set of VMs with various types, and the VMs are allocated to PMs. In the remaining of this paper, we use *Resource Allocation in*

L. Paquete and C. Zarges (Eds.): EvoCOP 2020, LNCS 12102, pp. 180–196, 2020.
https://doi.org/10.1007/978-3-030-43680-3_12

Container-based clouds (RAC) to represent the consolidation problem. In terms of difficulty, the two levels of allocation are both vector bin packing problems which are NP-hard [22]. Moreover, resource allocation in the first level, e.g., VM type selection, impacts the resource allocation in the second level.

Since it is impossible to find the optimal solution for a large scale *RAC* problem (e.g. over 1000 containers), existing studies mainly apply rule-based heuristics [6,11,14,22], and meta-heuristic algorithms [2,4,19] to find near-optimal solutions. Rule-based heuristics are greedy so they prone to stuck at local optimal solutions and they perform differently when facing various settings of VM types from multiple cloud providers. The meta-heuristics are promising algorithms. However, the current research either focuses on the problem of allocating containers directly to PMs or uses indirect representation which is inefficient in the searching process.

Group Genetic Algorithm (GGA) was proposed by Falkenauer [3] and has inspired many studies in solving the VM allocation problem [10,20]. Different from the standard GA, *GGA* applies a variable length of chromosome and domain-specific genetic operators such as inversion and rearrangement. *GGA* is designed for bin packing problem and uses a direct representation which avoids a decoding process. However, *GGAs* [3,16] can only solve one-level problems.

This research aims at proposing a novel *GGA* for the *RAC* problem to minimize the energy consumption. The proposed *GGA* provides the functionality of selecting VM types. Also, it has a direct representation and problem-specific operators to address the limitations of the dual-chromosome GA. To achieve our aim, we set up the following objectives:

1. To propose a new representation for the *RAC* problem;
2. To develop new genetic operators including gene-level crossover, unpack, rearrangement, and merge;
3. To evaluate our proposed approach by comparing it with the state-of-the-art algorithms: Rule-based (FF&BF/FF) approach [22] and two variations of dual-chromosome GAs [19].

The paper is organized as follows. Section 2 gives a background of our methodology and discusses related studies of the *RAC* problem. Section 3 presents the model of the problem. Then, Sect. 4 describes the proposed *GGA*. Section 5 illustrates the experiment design, results, and analysis. Section 6 summarizes the contributions and discusses the future works.

2   Related Work and Background

This section first reviews related works of the resource allocation in container-based clouds. Then, we provide a brief background of *GGA* [3].

2.1   Related Works

Current studies solve the *RAC* problem with two types of approaches, rule-based approaches, and meta-heuristics approaches. Piraghaj [14], Kaur [6], Mann [12],

Liu [9] and Zhang [22] treat the problem as a dynamic problem and propose AnyFit-based (e.g. First-Fit, Best-Fit) approaches to solve the problem. The proposed rules evaluate the candidate VMs and VM types to decide which VM to choose or which VM type to create. Overall, from the problem's perspective, as Wolke et al. [21] suggest that dynamic approaches are useful in some scenarios such as container migration and inferior in other scenarios such as initial container allocation. From the methods' perspective, the rules have a poor generality. Their performance varies when applying them to different settings of VM types (see Sect. 5.4). Another drawback is that these greedy rules are easily stuck at local optimal solutions.

A few meta-heuristics have been proposed, but they either focus on one-level allocation problem such as [4, 8], or uses an indirect representation [2, 19]. Guerrero et al. [4] propose an NSGA-II-based approach for a four-objective allocation problem. Lin et al. [8] propose an ant colony algorithm-based approach for the problem. In their models, containers are allocated directly to PMs without considering VMs. Tan et al. [2, 19] propose two meta-heuristic approaches for the RAC problem, an NSGA-II-based and a dual-chromosome GA (DGA). These approaches use indirect representations and they require a decoding process to interpret the representation to a solution. Overall, these algorithms search in the genotype space.

The current meta-heuristics have two shortcomings. The first drawback is that they [4, 8] only consider the one-level structure which inherently leads to local optimal solutions. The second drawback is that the decoding process of [19] can easily break the solutions (good combination of containers and VMs) from the previous generation. Therefore, it is hard to perform a directed search. As a consequence, the algorithms with indirect representation cannot find local optimal solutions efficiently.

Therefore, because of these drawbacks in the literature, we propose a meta-heuristic with a direct representation to solve the two-level RAC problem. The next section discusses the background of the GGA and explains how it can be adapted to our problem and meets our goal.

2.2   Group Genetic Algorithm (GGA)

GGA was proposed by Falkenauer [3] to solve the bin packing problem. GGA overcomes a major defect, the redundant encoding problem, in the ordering GA [15]. The ordering GA uses an encoded representation and the decoding process highly relies on items rather than the numbering of groups. For example, using two letters A and B to represent distinct groups, AAB and BBA are two solutions. However, in terms of grouping, these two solutions have the same meaning – the first two items are in the same group and the third item is in another group. To solve the redundant problem, GGA proposes a variable-length representation. The new crossover, mutation, and inversion operators directly operate on groups instead of items. Later on, Quiroz-Castellanos [16] embeds heuristics into the algorithm to speed up the search procedures.

GGA has been successfully applied to solve many bin packing problems such as ordering batch problems in warehouse [7], VM placement problem [5,10], and assembly line balancing problem [17]. However, it has not been used to solve any two-level vector bin packing problems. Our *RAC* problem is a two-level vector bin packing problem. It is promising to adopt *GGA*'s framework and propose problem-specific operators to solve our problem.

3 Problem Model

Resource Allocation in Container-based clouds (RAC) is a task of allocating a set of containers to a set of VMs of various types, then allocating the created VMs to a set of PMs. *VM selection* chooses an existing VM to allocate a container. *VM creation* selects a type of VM, creates a VM with the selected type and allocates the container to the new VM. The types of VM are defined by cloud providers. *PM selection* chooses an existing PM to allocate the new VM. If there is no available PM, a new PM will be created and the data center automatically allocates the new VM to the new PM. Since the PMs are homogeneous, no decision is needed for PM creation.

In the static setting of *RAC* problem, a set of containers $\mathcal{C} = \{c_1, \ldots, c_n\}$ arrives to the cloud to be allocated. Each container c_i has a CPU occupation $\zeta^{cpu}(c_i)$, a memory occupation $\zeta^{mem}(c_i)$. There is a set of VM types $\Gamma = \{\tau_1, \ldots, \tau_m\}$ that can be selected to allocate the containers. Each VM type τ_j has a CPU capacity $\Omega^{cpu}(\tau_j)$ and a memory capacity $\Omega^{mem}(\tau_j)$. Also, it has a CPU overhead $\pi^{cpu}(\tau_j)$ and memory overhead $\pi^{mem}(\tau_j)$, indicating the CPU and memory occupation for creating a new VM of that type. There is an unlimited set of PMs $\mathcal{P} = \{p_1, \ldots, \}$ for allocating the created VMs. Each PM p_k has a CPU capacity $\Omega^{cpu}(p_k)$ and a memory capacity $\Omega^{mem}(p_k)$.

The static *RAC* problem is subject to the following constraints:

1. Each container is allocated to one VM.
2. Each created VM is allocated to one PM.
3. For each created VM, the total CPU and memory occupations of the containers allocated to that VM does not exceed the corresponding VM capacity.
4. For each PM, the sum of the CPU and memory capacities of the VMs allocated on the PM does not exceed the corresponding PM's capacity.

The energy consumption is calculated as follows:

$$E = \sum_{k=1}^{K} E_k, \tag{1}$$

where E_k is the energy consumption of the kth PM (K is the number of PM used).

E_k is calculated as follows:

$$E_k = E_k^{idle} + (E_k^{full} - E_k^{idle}) \cdot \mu_k^{cpu}, \tag{2}$$

where E_k^{idle} and E_k^{full} indicate the energy consumption of the kth PM per time unit if it is idle and fully loaded, respectively. μ_k^{cpu} indicates the CPU utilization level of the kth PM. μ_k^{cpu} is calculated as follows.

$$\mu_k^{cpu} = \frac{\sum_{l=1}^{L}\left(\sum_{j=1}^{m}\pi^{cpu}(\tau_j)\cdot z_{jl} + \sum_{i=1}^{n}\Omega^{cpu}(c_i)\cdot x_{il}\right)\cdot y_{lk}}{\Omega^{cpu}(p_k)}, \tag{3}$$

where x_{il}, y_{lk} and z_{jl} are binary decision variables, and L is the number of created VMs. x_{il} takes 1 if c_i is allocated to the lth created VM, and 0 otherwise. y_{lk} takes 1 if the lth created VM is allocated to the kth PM, and 0 otherwise. z_{jl} takes 1 if the lth created VM is of type j, and 0 otherwise.

The static RAC problem is to find resource allocation with minimal overall energy consumption as shown as follows.

$$\min \sum_{k=1}^{K} E_k, \tag{4}$$

$$s.t. \sum_{l=1}^{L} x_{il} = 1, \ \forall \ i = 1,\ldots,n, \tag{5}$$

$$\sum_{k=1}^{K} y_{lk} = 1, \ \forall \ l = 1,\ldots,L, \tag{6}$$

$$\sum_{j=1}^{m} z_{jl} = 1, \ \forall \ l = 1,\ldots,L, \tag{7}$$

$$\sum_{i=1}^{n} \zeta^{res}(c_i)x_{il} \leq \sum_{j=1}^{m} \Omega^{res}(\tau_j)z_{jl},$$
$$\forall \ l = 1,\ldots,L, \ res \in \{cpu,mem\}, \tag{8}$$

$$\sum_{l=1}^{L}\sum_{j=1}^{m} \Omega^{res}(\tau_j)z_{jl} \leq \Omega^{res}(p_k),$$
$$\forall \ k = 1,\ldots,K, \ res \in \{cpu,mem\}, \tag{9}$$

$$x_{il}, y_{lk}, z_{jl} \in \{0,1\}, \tag{10}$$

where constraints (5) and (6) indicate that each container (or new created VM) is allocated to exactly one created VM (or PM). Constraint (7) indicates that each created VM must belong to a type. Constraint (8) implies that the total occupation of all the containers allocated to each created VM does not exceed its corresponding capacity. Constraint (9) indicates that the total capacity of the created VMs allocated to each PM does not exceed its corresponding capacity. Constraint (10) defines the domain of the decision variables.

The energy calculation (see Eq. 1) will be used as the fitness function of our proposed algorithm. The constraints of the model are used in the algorithm to ensure the solutions are valid.

4   The Proposed Group GA for the RAC Problem

This section describes our *GGA* approach for the *RAC* problem which includes a group representation and three problem specific operators.

4.1   Overall Framework

Algorithm 1 starts with the initialization of a population. The individual is represented as a list of PMs. Then, the algorithm enters a loop of evolutions where each loop is called a generation. In each generation, individuals are evaluated with a fitness function (Eq.(1)). For this algorithm, the top individuals are preserved and copied to the new population with Elitism [1]. Tournament selection [13] is used to direct the population to the high-fitness region. Then, we propose three problem-specific operators, gene-wise crossover, unpack, and merge. These operators modify the individuals so that they can perform an effective search in the solution space. Details of the three operators will be presented later.

Algorithm 1. Group genetic algorithm for the *RAC* problem

    Input   :   a set of containers, a set of VM types, a list of PMs,
    Output: an allocation of containers
1 population ← Initiailization;
2 *gen*; **for** *gen does not reach the maximum generation* **do**
3   fitness evaluation(population);
4   new population ← elitism(population);
5   **while** *has not fill the new population* **do**
6   parents ← tournament selection(population);
7   children ← gene-level crossover(parents);
8   unpack(children);
9   merge(children);
10   add children to the new population
11   **end**
12   *gen* ← *gen* + 1;
13 **end**
14 return an allocation of containers;

4.2   Representation

We use an individual (see Fig. 1) to represent a complete solution for a *RAC* problem. A individual consists of a list of PMs. Each PM consists of a list of VMs and each VM has a list of containers. This representation can be directly evaluated without using any decoding process. More importantly, the direct representation can be modified by heuristics at a specific point, e.g. switch two containers' allocation, without changing the structure of the entire solution.

Therefore, the disadvantage of indirect representation in *dual-chromosome GA* [19] can be avoided.

4.3 Initialization

The design of initialization aims at producing a diverse population of solutions.
For each individual, we first randomly
generate a permutation of containers.
Then, we allocate containers to VMs using
the First-Fit heuristic. If there is no VM
available, we create a VM with a random
type. Lastly, a list of VMs is allocated
to PMs with the First-Fit heuristic. This
representation ensures a diverse combination of containers and VMs. It also locates
the solutions in a relatively high-quality
region with First-Fit instead of Next-Fit.
This is because Next-Fit does not guarantee that a VM or a PM is filled while

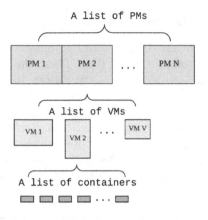

Fig. 1. Representation

First-Fit guarantees that. Therefore, the average quality obtained by First-Fit
is much better than Next-Fit.

4.4 Gene-Level Crossover

To inherit the useful parts from parents, one must define what is a "good gene".
In the bin packing problem, a good gene is at bins' level where well-filled bins can
lead to fewer bins [16]. Similarly, highly utilized PMs could lead to fewer PMs in
the allocation problem. Therefore, good gene as a PM with high utilization. In
our case, we apply the crossover twice according to the utilization of CPU and
memory respectively and generate two children.

The gene-level crossover preserves the highly utilized PMs from both parents.
In the beginning, we sort the PMs in both parents according to PMs' utilization
of CPU or memory in descending order. Then, the crossover compares the PMs
from two parents pairwisely by utilization (see Fig. 2). The winner's PM of the
pair will be preserved. Preservation includes three steps. First, the crossover
copies the VMs combination inside the PM including the types and number of
VMs. Second, the crossover checks whether a container from the original VM
has been allocated in the previous PMs. If the container has been allocated, then
the container will not be allocated again. In the end, some containers may not
be allocated to PMs. They are called *free containers*. These free containers are
reallocated with an operator called *rearrangement* which will be introduced in
the next section. After all the containers have been allocated, empty PMs and
VMs are removed from an individual.

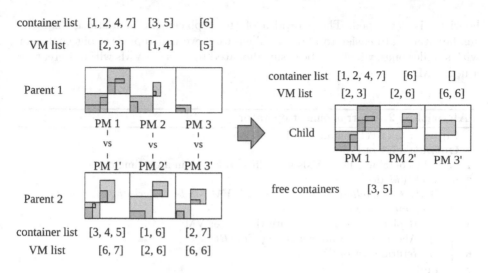

Fig. 2. Gene-level crossover

An example of the gene-level crossover is shown in Fig. 2. We first sort the PMs from parents according to their CPU utilization. Then, we compare PMs and preserve the structure of *PM 1*, *PM 2'*, and *PM 3'*. The containers in *PM 1* are preserved while the duplicated containers in *PM 2'* and *PM 3'* are removed. In the end, containers 3 and 5 become free containers and they will be allocated to these PMs using the *rearrangement* operator.

4.5 Rearrangement

Rearrangement inserts free items to bins. In the beginning (see Algorithm 2), we sort the containers according to the product of their normalized resources (see Eq. 11) in ascending order. Then, we check that in each VM, whether the smallest two containers can be replaced by the target container. If so, we replace the small containers with the target container. Otherwise, check the next VM. After replacing, we have two smaller containers need to be allocated. At this point, we apply *First-Fit (FF) & Random Creation (RC) / First-Fit (FF)* heuristics to allocate them. The *FF&RC/FF* heuristic uses FF to allocate containers to existing VMs. If no VM is available, we randomly create a new VM and allocate containers to it. Then, we use FF to allocate the new VM to PMs.

$$R = \frac{\zeta^{cpu}(c_i)}{\Omega^{cpu}(p_k)} \cdot \frac{\zeta^{mem}(c_i)}{\Omega^{mem}(p_k)} \tag{11}$$

Our rearrangement operator is inspired by [16] to avoid the drawback of First-Fit (FF) and further improve the allocation of a VM. In the bin packing problem, FF-based approaches [3,16] have been widely used. However, a simple FF-based approach cannot change the existing packing of a bin. Hence, a replacement heuristic is developed. The core idea of the replacement heuristic is that the smaller items are easier to allocate. Therefore, we can replace a big container with smaller ones, which can be easily allocated to existing VMs without creating a new VM.

Algorithm 2. Rearrangement operator

  Input : a target container, a list of PMs,
  Output: a list of PMs
1 Sort the containers in all VMs according to Eq. 11 in ascending order;
2 **for** *each VM* **do**
3   **if** *the two smallest containers in each VM can be replaced by the target container* **then**
4   Replace two containers with the target VM;
5   Allocate two containers using *FF&RC/FF*;
6   return a list of PMs;
7   **end**
8 **end**
9 Allocate the target container using *FF&RC/FF*;
10 return a list of PMs;

4.6 Unpack

Unpack operator eliminates low-utilized PMs and reallocates their containers. This operator prevents premature convergence and introduces new gene components into the current population.

The operator has two steps. First, it calculates the probability of unpacking a PM according to Eq. (12). The lower CPU utilization of a PM, the higher chance it will be unpacked. Second, it unpacks PMs in a roulette wheel style. After unpacking, the free containers are reallocated with the *rearrangement* operator.

$$probability = \frac{1 - \Omega^{cpu}(p_k)}{\sum_{k=1}^{K} 1 - \Omega^{cpu}(p_k)} \tag{12}$$

The unpack operator is adaptive with the evolution process. In the beginning, the average utilization of PMs is low, therefore, more PMs are unpacked. As the population evolved through some generation, highest utilized PMs move to the head of an individual and have a low chance to be unpacked. Therefore, the good genes are preserved and new genes are introduced by the *rearrangement* operator.

4.7 Merge

The merge operator replaces small VMs with a bigger one to reduce the *free resources* in PMs. Free resources here refer the resources that have not been allocated to any VMs. The merge operator can improve the utilization of PM by reducing the free resources in PMs as well as the overheads from VMs.

Merge operators have two alternative functionalities, merge and enlarge. In the first one, it goes through all the PMs and checks whether the two smallest VMs can be replaced by a larger VM type. If it is possible, all the containers are migrated from these two small VMs to the new larger VM and the small VMs are removed. If we cannot replace two VMs with a larger one, we attempt to replace the smallest VM with a larger one for which a large VM type is also selected randomly.

5   Experiment

The overall goal of the experiment is to test the performance of our proposed *GGA* in terms of energy consumption. We conduct experiments on a real-world dataset and compare the results with three benchmark algorithms (a rule-based approach *FF&BF/FF* and two variations of the *dual-chromosome GA*). Then, we analyze the performance of these approaches and explain the pros and cons of them. Details are shown below.

5.1   Dataset and Test Instance

We design 8 test instances (see Table 1) which contain an increasing number of containers (from 200 to 1500) and two sets of VM types. We use a real-world application trace (AuverGrid trace [18]) as the resource requirements of containers. To generate the containers' resource requirements, we select the first 400,000 lines of the trace from the original datasets. Then we filtered the trace to exclude the containers that require more resources than the largest VM. The last step randomly samples a set of resource requirements and use them to define the containers to be allocated.

Table 1. Test instances

Instance	VM types	Number of containers	Instance	VM types	Number of containers
1	Synthetic VM types	200	5	Real-world VM types	200
2	Synthetic VM types	500	6	Real-world VM types	500
3	Synthetic VM types	1000	7	Real-world VM types	1000
4	Synthetic VM types	1500	8	Real-world VM types	1500

For the settings of PMs and VMs, we assume homogeneous PMs which have 8 cores and a total capacity of [13200 MHz, 16000 MB]. The maximum energy

Table 2. VM types

Real world VM types							
VM types	[CPU, Memory]	VM types	[CPU, Memory]	VM types	[CPU, Memory]	VM types	[CPU, Memory]
1	[206.25, 250]	6	[412.5, 1000]	11	[825, 2000]	16	[825, 1875]
2	[412.5, 500]	7	[825, 4000]	12	[1650, 250]	17	[1650, 3750]
3	[825, 1000]	8	[206.25, 500]	13	[1650, 500]	18	[412.5, 1312.5]
4	[1650, 2000]	9	[412.5, 2000]	14	[1650, 1000]	19	[825, 2625]
5	[412.5, 250]	10	[412.5, 4000]	15	[412.5, 937.5]	20	[2475, 2625]
Synthetic VM types							
1	[719, 2005]	4	[1135, 3542]	7	[1363, 2634]	10	[2100, 3013]
2	[917, 951]	5	[1231, 1989]	8	[1648, 1538]		
3	[1032, 1009]	6	[1311, 3238]	9	[2047, 1181]		

consumption for the PM is set to 540 KWh the same setting as [11]. We design two sets of VM types (see Table 2), a real-world VMs (20 types from Amazon EC2) and a synthetic set of VMs (10 types). The real-world VM types are proportional whereas the synthetic ones are random. The values of CPU and memory of synthetic VM types are sampled from [0, 3300 MHz] and [0, 4000 MB] representing the capacity of one core.

5.2 Benchmark Algorithms

FF&BF/FF [11,22] uses three heuristics to allocate containers. It uses First-Fit heuristics to allocate both containers and VMs and applies a Best Fit (BF) for selecting VM types. Whenever no existing VM can host a given container, the BF selects a type of VM which has just enough resource to host the container. Explicitly, BF selects the VM which has the minimum normalized free resources according to Eq. 13.

$$Free\ resources = min\{\frac{\Omega^{cpu}(\tau_j) - \zeta^{cpu}(c_i) - \pi^{cpu}(\tau_j)}{\Omega^{cpu}(p_k)} \text{ and } \frac{\Omega^{mem}(\tau_j) - \zeta^{mem}(c_i) - \pi^{mem}(\tau_j)}{\Omega^{mem}(p_k)}\}$$
(13)

Dual-chromosome GA is a recent approach proposed in [19] to solve the resource allocation problem in container-based clouds. This approach uses a dual chromosome representation which includes two vectors, one represents a permutation of containers, the other represents the selected VM types. An individual requires a decoding process to construct the dual-chromosome into a solution. The rest of the algorithm follows a standard GA process with vector-based crossover and mutation operators.

This paper compares with two variations of the *dual-chromosome GA* with two decoding processes. The original work [19] applies a Next-Fit (NF) decoding. We refer it as *DGA-NF* in the following content. We implement a different version that applies a First-Fit (FF) decoding called *DGA-FF*.

Table 3. Parameter settings

Parameter	Description
Runs	30
Crossover	70%
Mutation rate for dual-chromosome GA	10%
Elitism	Top 5 individuals
Stopping criteria	12 s
Population	100
Selection	Tournament selection (size = 7)

In the experiments, we also compare the wasted resources in the allocation. The wasted resources include all the free resources in both VMs and PMs as well as the overheads used by VMs (see Eq. 14).

$$wasted\ resources = min\{\frac{\Omega^{cpu}(p_k) - \sum_{i=1}^{n} \zeta^{cpu}(c_i) \cdot x_{il}}{\Omega^{cpu}(p_k)} \ and \ \frac{\Omega^{mem}(p_k) - \sum_{i=1}^{n} \zeta^{mem}(c_i) \cdot x_{il}}{\Omega^{mem}(p_k)}\}$$

$$(14)$$

5.3 Parameter Settings

The parameter setting of GGA and two dual-chromosome GAs are listed in Table 3. In addition to the operators that we proposed, we apply Elitism with size 5 and tournament selection with size 7. To ensure that all algorithms have the same computation time, we set the stopping criteria of all GAs to 12 s (all algorithms finished in this period of time).

All algorithms were implemented in Java version 8 and the experiments were conducted on i7-4790 3.6 GHz with 8 GB of RAM running Linux Arch 4.14.15. We applied the Wilcoxon rank-sum to test the statistic significance.

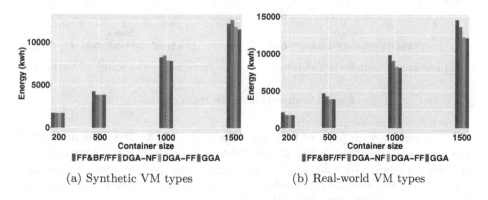

(a) Synthetic VM types   (b) Real-world VM types

Fig. 3. Comparison of the average energy consumption

Table 4. Mean and standard deviation of the test instances with 95% confident interval.

Synthetic VM types				
	200	500	1000	1500
FF&BF/FF	1708.0 ± 0	4244.2 ± 0	8259.5 ± 0	12176.0 ± 0
DGA-NF	1685.6 ± 0.3	3838.5 ± 1.1	8,485.3 ± 94.1	12,625.8 ± 50.6
DGA-FF	1684.6 ± 0.2	3758.4 ± 151.1	7,865.7657 ± 1.0	11,795.3957 ± 1.5
GGA	1686.0 ± 0.1	3571.4 ± 177.1	7,833.9 ± 41.1	11,490.7 ± 108.8

Real-world VM types				
	200	500	1000	1500
FF&BF/FF	2093.2 ± 0	4635.0 ± 0	9809.2 ± 0	14500.4 ± 0
DGA-NF	1683.6 ± 0.4	4213.1 ± 1.8	9,027.9 ± 3.1	13,580.9 ± 93.9
DGA-FF	1682.3 ± 0.2	3827.8 ± 1.3	8,222.3681 ± 40.7	12,180.0944 ± 1.9
GGA	1683.1 ± 0.5	3828.2 ± 2.3	8,091.7 ± 91.1	12,083.8 ± 51.7

5.4   Results

This section illustrates the performance comparison among the four algorithms in terms of energy consumption. Then, we explain the drawbacks of the compared algorithms by comparing the convergence, the number of VMs and the wasted resources in the allocation. Lastly, we compare the execution time of the four algorithms.

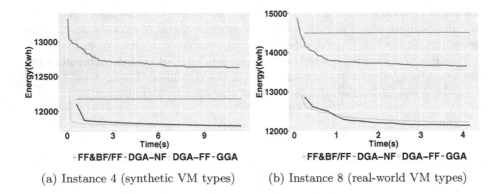

(a) Instance 4 (synthetic VM types)   (b) Instance 8 (real-world VM types)

Fig. 4. Comparison of the convergence

The energy consumption of four algorithms running for the same amount of time (12 s) are compared in Fig. 3 and Table.4. This ensures the comparison is fair. Our proposed *GGA* approach consistently achieves the best performance than the *FF&BF/FF* and two *dual-chromosome GA* approaches, *DGA-NF* and *DGA-FF*, in large instances. The *DGA-FF* has a similar performance with *GGA* in the small instance (less than 1500 containers) but it performs poorly in the large instances. The *DGA-NF* performs better than *FF&BF/FF* in most of the instances except instance 3 and 4 (1000 and 1500 containers with synthetic VM types). In instances with 200 and 500 containers, *DGA-FF* and *GGA* have similar performances. In larger instances, *GGA* has clearly show its advantages.

Due to the space limit, we show in Fig. 4 the convergence curves in terms of computation time from instance 4 and 8. In most instances except instance 3 and 4, the convergence curves are similar to instance 8 where we observe the *FF&BF/FF* shows a flat line because it has no searching process. *FF&BF/FF* is also easily affected by the set of available VM types as it performs well in the synthetic VM types and performs poorly in the real-world data set. The *DGA-NF* starts with a much higher energy consumption than other algorithms. Although *DGA-NF* reaches convergence, its final fitness value cannot compete with the initial fitness from *DGA-FF* and *GGA*. In instance 3 and 4, the *DGA-NF* worse than the *FF&BF/FF* approach. *DGA-FF* and *GGA* have a similar starting point. In instance 4, *DGA-FF* and *GGA* have a similar pattern while *GGA* outperforms in instance 8 after 1 s.

The major defect of *DGA-NF* is the decoding process. Compared to FF, NF closes a bin (such as VM and PM) whenever the current item (such as container and VM) cannot allocate to it while FF never closes a bin so that the future items can be still put into the unfilled bins. It means that NF cannot guarantee a VM is filled with containers. Consequently, we may observe *DGA-NF* starts from a bad allocation and takes a long time to converge. Even though replacing NF with FF can improve the performance of *DGA*. However, the *DGA-FF* is still inferior to the *GGA* approach.

The number of VMs (left-hand side) and the wasted resources (right-hand side) are compared in Fig. 5. The *FF&BF/FF* always uses the greatest number of VMs and has the highest wasted resources. For most instances, the *dual-chromosome* algorithms use fewer VMs and have fewer wasted resources except in instance 4. Our proposed *GGA* always uses the least number of VMs and has the least wasted resources.

Due to the overheads and resource segmentation, the number of VMs is generally proportional to the wasted resources. The *FF&BF/FF* always creates a VM that has the least resources to host a container, and therefore, creates a large number of small VMs. *DGA-NF* has a high wasted resource in instance 4 because *DGA-NF* cannot fill VMs with container, hence, there are more free resources in VMs and PMs than the overheads of VMs. *DGA-FF* and *GGA* use fewer VMs. However, *DGA-FF* does not have the mechanism to reduce the number of VMs.

On the other hand, among all the algorithms, *GGA* can generate allocation solutions with the least wasted resources thanks to the merge operator. Without deliberately merging smaller VMs into larger ones, a PM could be filled with a large number of small VMs.

In summary, our propose *GGA* can find an allocation that leads to the least energy consumption in all the test instances. The performance of *dual-chromosome GA* varies with the decoding process.

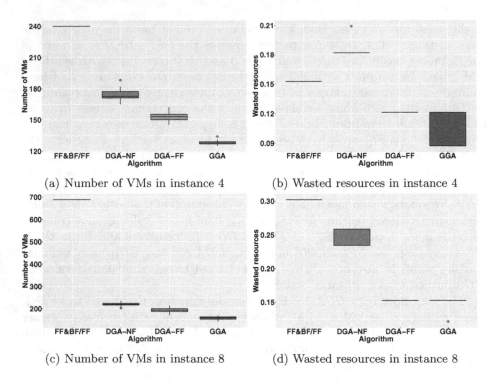

(a) Number of VMs in instance 4   (b) Wasted resources in instance 4

(c) Number of VMs in instance 8   (d) Wasted resources in instance 8

Fig. 5. Number of VMs and wastes in instance 4 and 8

6   Conclusion and Future Work

This work proposes a *Group GA (GGA)*-based approach to solve the resource allocation problem in container-based clouds. The experiments show that our proposed *GGA* approach outperforms three state-of-the-art approaches, a rule-based *FF&BF/FF* approach and two variations of *dual-chromosome GA* [19] in terms of energy consumption. For our GGA, we propose three novel problem-specific operators, gene-level crossover, rearrangement, and unpack. These operators have shown effectiveness in searching good combinations of containers and VM types. Also, these operators can effectively search for better solutions directly on the representation. Current operators have a high computation cost in each generation. In the future, we will focus on improving the efficiency by applying clustering-based preprocessing approaches.

References

1. Bhandari, D., Murthy, C., Pal, S.K.: Genetic algorithm with elitist model and its convergence. Int. J. Pattern Recogn. Artif. Intell. **10**(06), 731–747 (1996)
2. Tan, B., Ma, H., Mei, Y.: A NSGA-II-based approach for service resource allocation in Cloud. In: IEEE Congress on Evolutionary Computation (CEC), pp. 2574–2581 (2017)

3. Falkenauer, E.: A hybrid grouping genetic algorithm for bin packing. J. Heuristics **2**(1), 5–30 (1996)
4. Guerrero, C., Lera, I., Juiz, C.: Genetic algorithm for multi-objective optimization of container allocation in cloud architecture. J. Grid Comput. **16**(1), 113–135 (2018)
5. Kaaouache, M.A., Bouamama, S.: Solving bin packing problem with a hybrid genetic algorithm for VM placement in cloud. Proc. Comput. Sci. **60**, 1061–1069 (2015)
6. Kaur, K., Dhand, T., Kumar, N., Zeadally, S.: Container-as-a-service at the edge: trade-off between energy efficiency and service availability at fog nano data centers. IEEE Wirel. Commun. **24**(3), 48–56 (2017)
7. Koch, S., Wäscher, G.: A grouping genetic algorithm for the order batching problem in distribution warehouses. J. Bus. Econ. **86**(1–2), 131–153 (2016)
8. Lin, M., Xi, J., Bai, W., Wu, J.: Ant colony algorithm for multi-objective optimization of container-based microservice scheduling in cloud. IEEE Access **7**, 83088–83100 (2019)
9. Liu, B., Li, P., Lin, W., Shu, N., Li, Y., Chang, V.: A new container scheduling algorithm based on multi-objective optimization. Soft Comput. **22**(23), 7741–7752 (2018). https://doi.org/10.1007/s00500-018-3403-7
10. Liu, X.F., Zhan, Z.H., Deng, J.D., Li, Y., Gu, T., Zhang, J.: An energy efficient ant colony system for virtual machine placement in cloud computing. IEEE Trans. Evol. Comput. **22**(1), 113–128 (2016)
11. Mann, Z.Á.: Interplay of virtual machine selection and virtual machine placement. In: Aiello, M., Johnsen, E.B., Dustdar, S., Georgievski, I. (eds.) ESOCC 2016. LNCS, vol. 9846, pp. 137–151. Springer, Cham (2016). https://doi.org/10.1007/978-3-319-44482-6_9
12. Mann, Z.Á.: Resource optimization across the cloud stack. IEEE Trans. Parallel Distrib. Syst. **29**(1), 169–182 (2018)
13. Miller, B.L., Goldberg, D.E.: Genetic algorithms, tournament selection, and the effects of noise. Complex Syst. **9**(3), 193–212 (1995)
14. Piraghaj, S.F., Dastjerdi, A.V., Calheiros, R.N., Buyya, R.: A framework and algorithm for energy efficient container consolidation in cloud data centers. In: International Conference on Data Science and Data Intensive Systems, pp. 368–375. IEEE (2015)
15. Poon, P.W., Carter, J.N.: Genetic algorithm crossover operators for ordering applications. Comput. Oper. Res. **22**(1), 135–147 (1995)
16. Quiroz-Castellanos, M., Cruz-Reyes, L., Torres-Jimenez, J., Gómez, C., Huacuja, H.J.F., Alvim, A.C.: A grouping genetic algorithm with controlled gene transmission for the bin packing problem. Comput. Oper. Res. **55**, 52–64 (2015)
17. Şahin, M., Kellegöz, T.: An efficient grouping genetic algorithm for u-shaped assembly line balancing problems with maximizing production rate. Memetic Comput. **9**(3), 213–229 (2017)
18. Shen, S., van Beek, V., Iosup, A.: Statistical characterization of business-critical workloads hosted in cloud datacenters. In: IEEE/ACM International Symposium on Cluster, Cloud and Grid Computing, pp. 465–474. IEEE (2015)
19. Tan, B., Ma, H., Mei, Y.: Novel genetic algorithm with dual chromosome representation for resource allocation in container-based clouds. In: International Conference on Cloud Computing, pp. 452–456. IEEE (2019)
20. Wen, Y., Li, Z., Jin, S., Lin, C., Liu, Z.: Energy-efficient virtual resource dynamic integration method in cloud computing. IEEE Access **5**, 12214–12223 (2017)

21. Wolke, A., Bichler, M., Setzer, T.: Planning vs. dynamic control: resource allocation in corporate clouds. IEEE Trans. Cloud Comput. 4(3), 322–335 (2016)
22. Zhang, R., Zhong, A., Dong, B., Tian, F., Li, R.: Container-VM-PM architecture: a novel architecture for docker container placement. In: Luo, M., Zhang, L.-J. (eds.) CLOUD 2018. LNCS, vol. 10967, pp. 128–140. Springer, Cham (2018). https://doi.org/10.1007/978-3-319-94295-7_9

The Local Optima Level
in Chemotherapy Schedule Optimisation

Sarah L. Thomson$^{(\boxtimes)}$ (iD) and Gabriela Ochoa (iD)

Computing Science and Mathematics, University of Stirling, Stirling, UK
{s.l.thomson,gabriela.ochoa}@stir.ac.uk

Abstract. In this paper a multi-drug Chemotherapy Schedule Optimisation Problem (CSOP) is subject to Local Optima Network (LON) analysis. LONs capture global patterns in fitness landscapes. CSOPs have not previously been subject to fitness landscape analysis. We fill this gap: LONs are constructed and studied for meaningful structure. The CSOP formulation presents novel challenges and questions for the LON model because there are infeasible regions in the fitness landscape and an unknown global optimum; it also brings a topic from healthcare to LON analysis. Two LON Construction algorithms are proposed for sampling CSOP fitness landscapes: a Markov-Chain Construction Algorithm and a Hybrid Construction Algorithm. The results provide new insight into LONs of highly-constrained spaces, and into the proficiency of search operators on the CSOP. Iterated Local Search and Memetic Search, which are the foundations for the LON algorithms, are found to markedly out-perform a Genetic Algorithm from the literature.

Keywords: Combinatorial fitness landscapes · Local Optima Networks · Search space analysis

1 Introduction

Local Optima Networks (LONs) [1] are used to study fitness landscapes. Analysis of them provides insight into how optimisation problems and search algorithms interact together. LONs capture *global* patterns at the Local Optima Level (LOL) in landscapes and have mostly been extracted for benchmark combinatorial optimisation problems such as NK Landscapes [1–3], the Quadratic Assignment Problem (QAP) [4–6], and the Travelling Salesman Problem (TSP) [7–9].

Studies in non-benchmark problem domains have been sparse and have consisted of computational protein modelling [10] and feature selection [11]. These were steps towards bringing LON analysis to unmapped real-world problems. This type of case study, demonstrating LON efficacy, is needed for convincing possible industry collaborators. Large and highly-constrained problems should ideally be used in case studies (by 'large' we mean hundreds of dimensions), in pursuit of simulating environments typical of real-world optimisation problems.

© Springer Nature Switzerland AG 2020
L. Paquete and C. Zarges (Eds.): EvoCOP 2020, LNCS 12102, pp. 197–213, 2020.
https://doi.org/10.1007/978-3-030-43680-3_13

Chemotherapy Schedule Optimisation Problems (CSOPs) [12] have been the subject of several research papers in evolutionary computation [13–17]. One instance was formulated to reflect real-life chemotherapy drug response closely [13] and the tumour shrinkage model used in the fitness function has been subject to extensive clinical testing [18]. The instance, alongside other CSOP formulations, has not been subject to fitness landscape analysis (although some authors have made passing remarks about CSOP landscapes [19,20]).

We conduct a first fitness landscape analysis on CSOP, focussing on the LOL with the use of LOLs. Two LON Construction algorithms are proposed—the first has Iterated Local Search (ILS) as its foundation; the second has Memetic Search (MS) as the foundation. LONs are then produced and their attributes and fitness distributions are compared. A study of the feasibility trajectories in the LONs is also presented. Later on, algorithm performance results suggest our search algorithms (ILS and MS) outperform a GA from the literature for the CSOP. In summary, the present work contributes in the following ways:

1. First fitness landscape analysis of CSOP, lending to new insights of the problem interacting with search operators;
2. The presence of infeasible solutions in the landscapes is new to LON research;
3. Two LON Construction algorithms for the CSOP are proposed (which can also be easily applied to an arbitrary binary-encoded problem).
4. Two search algorithms are offered which outperform a GA from the literature (ILS and MS; a separate MS has been used on a CSOP formulation before but with different fitness function, constraints, and solution encoding [19]).

1.1   Background

We use a multi-drug CSOP which was initially formulated and described in 1998 [13] and then further studied in later research [14–17,21]. As asserted in the original paper, a multi-drug CSOP can have a binary representation where each gene, i, is set iff a particular concentration of a particular drug (of number n) is administered at a particular time interval (t, from within defined time intervals). As suggested in the literature [21], we set the number of drugs $n = 10$ and the number of time intervals for doses, t, also at 10. There are four allowed concentrations for each drug, $p = 4$, giving each binary solution a length of 400, i.e. $n \times t \times p$. The number of possible solutions, and the size of the configuration space, is extremely large at 2^{ntp} i.e. 2^{400}.

1.2   Fitness Function

We consider *curative* chemotherapy treatment here, meaning tumour eradication is the aim. This is the primary (and only) objective. For this single-objective case, fitness is calculated with respect to the chemotherapy schedule *minimising* tumour size (in number of cells). This is done through *maximising* the combined effect of drugs in the schedule against the tumour. In considering the tumour's shrinkage response, a mathematical function is needed. The most popular model in the literature is called the *Gompertz Growth Model* [18], which has a linear cell-loss effect and has been validated by significant clinical experiments. The formula is given in Eq. 1:

$$\frac{dN}{dt} = N(t) \left(\lambda \ln \left(\frac{\theta}{N(t)} \right) - \sum_{j=1}^{d} k_j \sum_{n}^{i=1} C_{ij} \left\{ H(t - t_i) - H(t - t_{i+1}) \right\} \right) \quad (1)$$

with the components as follows: $N(t)$ is the cancerous cell count at time interval t; λ and θ are parameters pertaining to tumour growth; $H(t)$ is the Heaviside step function; k_j denotes the efficacy of chemotherapy drugs; and C_{ij} is the concentration levels of the drugs administered.

The actual fitness function is quite complex, including penalties based on feasibility distances, and in the interest of space we refer the interested reader to a comprehensive description [13] (pp. 106–107). In essence, initial fitness is calculated with respect to the total impact on the tumour for the treatment schedule. Individual impacts for each drug are known. The objective is to maximise the combined impact of all the drugs in the schedule (at the specified concentrations, and at the specified time-slots). The *maximisation* of this will *minimise* the tumour.

Following the drug impact fitness calculation, the solution is checked for constraint violations and the fitness is penalised accordingly (see [13] for details). Any violation will result in a fitness below zero. A feasible solution has fitness zero or above.

The constraints are as follows: the tumour is not allowed above a particular size; the maximum cumulative dose of drugs cannot exceeded the specified limits for each individual drug; and the limit on toxic chemotherapy side-effects cannot be exceeded (for each time interval). In all cases the magnitude of the violation is captured through proportional subtraction from the fitness sum.

Mathematically the fitness function is subject to these constraints:

1. Maximum allowable *cumulative* C_{cum} dosage for each drug:

$$g_1(c) = \left\{ C_{cum\,j} - \sum_{i=1}^{n} C_{ij} \geq 0 \,\vdots\, \forall j \in \overline{1,d} \right\} \quad (2)$$

2. Maximum allowable size of the tumour, i.e. number of cancerous cells, N:

$$g_2(c) = \left\{ N_{max} - N(t_i) \geq 0 \,\vdots\, \forall i \in \overline{1,n} \right\} \quad (3)$$

3. A threshold for the known toxic side-effects of using multiple drugs in chemotherapy treatment:

$$g_3(c) = \left\{ C_{s-e_k} - \sum_{j=1}^{d} \eta_{kj} C_{ij} \geq 0 \,\vdots\, \forall i \in \overline{1,n}, \forall k \in \overline{1,m} \right\} \quad (4)$$

In the constraint seen in Eq. 4, the variables η_{kj} are the known possibility of harming the kth organ (for example, the heart) through administering the jth drug.

1.3 Evolutionary Search Algorithms

Evolutionary algorithms have been used with success for CSOPs; in particular, Genetic Algorithms (GAs) have dominated [13,21–23], although other approaches have been utilised, such as Estimation of Distribution Algorithms [15,17]; Simulated Annealing variants [23,24]; Memetic Algorithm (MA) [19]; and Evolutionary Strategies [23,25]. A GA from the literature [17] is used as the foundation for the *Hybrid* LON Construction algorithm proposed here (detailed later in Sect. 2.2) and is also used later on in conducting optimisation on the problem to collect search difficulty information.

2   Methodology

This section describes the LON Construction algorithms proposed for studying CSOP fitness landscapes. Our aim is examining the topological features forming when optimisation search operators are moving on the CSOP configuration space. The particular focus is on *global-scale* local optima connectivity patterns.

2.1   Markov-Chain LON Construction Algorithm

To align with existing LON Construction algorithms for benchmark domains such as TSP and QAP [6,26,27] we instrument an algorithm using Iterated Local Search (ILS) as the vehicle. ILS is naturally suited to constructing LONs: each iteration identifies a transformation between local optima and this can straightforwardly be added as an edge to a LON. We refer to the ILS-driven LON Construction algorithm as *Markov-Chain* LON Construction—to avoid confusion, because ILS is also used later on to collect difficulty information about the CSOP.

Markov-Chain LON Construction tracks thirty independent ILS runs, which begin from random solutions. The local search is best-improvement and uses one-flip neighbourhood. Perturbation flips thirty bits. Improving local optima are always accepted; 10% of the time, worsening local optima are accepted too. All accepted local optima are added as LON nodes and the transformation is logged as a LON edge (if the edge exists already, the weight is incremented). Runs terminate after 1000 iterations. Parameters were chosen in response to observations about preliminary runs.

Nodes and edges from the thirty runs are joined together to form a single LON for the problem. Our initial intention was to mirror parameter choices in previous LON Construction works [27] but those choices were for much smaller search space sizes and the computation was therefore more feasible for their circumstance. The complete process for *Markov-Chain* LON Construction is provided in Algorithm 1.

Algorithm 1. Markov-Chain LON Construction

1: Search space S, Fitness function f,
2: Perturbation strength k, Stopping threshold t, Number of runs r
3: $runs \leftarrow 0$
4: **repeat**
5:     Choose initial random solution $s_0 \in S$
6:     $l \leftarrow$ LocalSearch(s_0)
7:     $i \leftarrow 0$
8:     **repeat**
9:         $s' \leftarrow$ Perturbation(l_1, k)
10:         $l_2 \leftarrow$ LocalSearch(s')
11:         **if** $f(l_2) \leq f(l_1)$ **then**
12:             $l_1 \leftarrow l_2$
13:         **end if**
14:         $LON = LON + \text{nodes}(l_1, l_2)$
15:         $LON = LON + \text{edge}(l_1 \longrightarrow l_2)$
16:         $i \leftarrow i + 1$
17:     **until** $i \geq t$ **return** l
18:     $runs \leftarrow r + 1$
19: **until** $runs \geq r$

2.2   Hybrid LON Construction Algorithm

Genetic Algorithms (GAs) have been successful in finding good approximate solutions in CSOP [16]; it follows that a GA is a reasonable foundation for CSOP LON Construction. When LON Construction algorithms share operators with successful heuristics, the constructed LONs should *infer* future landscapes that might be induced during genuine optimisation.

Our LON Construction algorithm originates from a generational GA from the literature [14]. By definition LONs contain only local optima. GAs, of course, do not guarantee local optima in the population, which necessitates the addition of local search to the algorithm, resulting in a Memetic Search (MS). The Memetic Search-driven LON tracking process is hereafter referred to by *Hybrid* LON Construction, to differentiate from a MS used later for collecting problem difficulty information.

The algorithmic process for creating the LON is as follows. The algorithm runs for 100 generations; at each generation the fittest 10% of offspring are subject to local search to produce local optima. To deem a node a local optimum, one-flip best-improvement hill-climbing is applied for 100 iterations. The nodes are added to the LON and are put into the next generation. The set of local optima are then deterministically recombined with one another. The offspring are possibly mutated according to the mutation rate, before being subject to local search. All four local optima (parent one and two, child one and two) are then added as nodes to the LON. Similarly, four edges are added to the network: from parent one to child one; parent one to child two; parent two to child one; and parent two to child two.

Algorithm 2. Hybrid LON Construction: Part 1

1: **procedure** HYBRID LON CONSTRUCTION(population size ps, generations g, percent fittest individuals pf, mutation probability mp, crossover probability cp, length evolution path lp, search space S, fitness function f)
2:　　$LON = \emptyset$　　　　　　　　　　　　　　　▷ local optima network
3:　　$P = \text{randomPopulation}(S, ps)$　　　　　　▷ Initial random population
4:　　$fit = \text{SelectFittest}(P, pf, f)$　　　　　　　▷ fittest individuals
5:　　$iterations = 0$　　　　　　　　▷ counter for generations completed
6:　　**repeat**
7:　　　　$P = \text{GENETICPROCESS}(P, ps)$
8:　　　　$fit = \text{SelectFittest}(P, pf)$
9:　　　　**for** $sol \in fit$ **do**
10:　　　　　　$sol = \text{HillClimb}(sol)$
11:　　　　**end for**
12:　　　　**for** $mom \in fit$ **do**
13:　　　　　　**for** $dad \in fit$ **do**
14:　　　　　　　　$child_1, child_2 = \text{MEMETICEVOLUTION}(mom, dad)$
15:　　　　　　　　$iterations = iterations + 1$
16:　　　　　　**end for**
17:　　　　**end for**
18:　　**until** $iterations \geq g$
19:　　**for** $mom \in LON$ **do**
20:　　　　**for** $dad \in LON$ **do**
21:　　　　　　$steps = 0$
22:　　　　　　**repeat**
23:　　　　　　　$child_1, child_2 = \text{MEMETICEVOLUTION}(mom, dad)$
24:　　　　　　　$mom, dad = child_1, child_2$
25:　　　　　　　$steps = steps + 1$
26:　　　　　　**until** $steps \geq lp$
27:　　　　**end for**
28:　　**end for**
29: **end procedure**

After all generations are complete, LON nodes undergo another evolutionary process. For each pairwise combination of nodes, the following is repeated ten times: the solutions are deterministically recombined with one another; the offspring are probabilistically subject to mutation; the offspring are subject to local search. They are added as LON nodes, and transformations from parent to child are added as edges. After this, the locally-optimised offspring become the parents for the next iteration of the same process. The steps repeat ten times. In this way, each pair of original LON nodes (from the 100-generation MS) are the ancestors in a ten-generation evolutionary trajectory. This was a deliberate design choice to facilitate LONs containing sequences of evolution for local optima. Without this, the LON would consist of many isolated pairs of nodes and would be difficult to study for meaningful structure. The complete process for the *Hybrid* LON Construction algorithm is shown in Algorithms 2 and 3.

Algorithm 3. Hybrid LON Construction: Part 2

1: **procedure** GENETICPROCESS(P, ps)
2:     **repeat**
3:         $mom, dad = $ Selection(P)
4:         **if** cp **then**
5:             $child_1, child_2 = $ Crossover(mom, dad)
6:         **end if**
7:         **if** mp **then**
8:             $child_1, child_2 = $ Mutation($child_1, child_2$)
9:         **end if**
10:        $P[mom] = child_1$
11:        $P[dad] = child_2$
12:     **until** iterations $\geq ps/2$
13: **end procedure**

1: **procedure** MEMETICEVOLUTION(mom, dad)
2:     $child_1, child_2 = $ Crossover(mom, dad)
3:     **if** mp **then**
4:         $child_1, child_2 = $ Mutation($child_1, child_2$)
5:     **end if**
6:     $child_1, child_2 = $ HillClimb($child_1, child_2$)
7:     $LON = LON + $ nodes($child_1, child_2$)
8:     $LON = LON + $ edges($\{mom \longrightarrow child_1\}$, $\{mom \longrightarrow child_2\}$, $\{dad \longrightarrow child_1\}$, $\{dad \longrightarrow child_2\}$)
9:     **return** $child_1, child_2$
10: **end procedure**

3 Visualisations

Visual analysis of LONs can provide an abstracted view of the Local Optima Level, which is a multi-dimensional complex system. Sometimes, patterns observable in visual analysis help to explain search algorithm performance on the associated combinatorial problem.

Markov-Chain LON Construction and Hybrid LON Construction algorithms produce networks with thousands of nodes. For meaningful visualisation, pruned sub-networks are constructed. The 'elite' nodes of the LONs are chosen for this. For the Markov-Chain LON, these are nodes in the top 2% of the fitness distribution. The Hybrid LON has more nodes, so only the top 0.05% are visualised. It follows that this lifts the veil on the most promising regions reached by the algorithms.

Figure 1 shows plots for two LONs of the same CSOP. The top Figure is the LON constructed by the Markov-Chain method; on the bottom was constructed with the Hybrid method. Edges encode sequences of search operations.

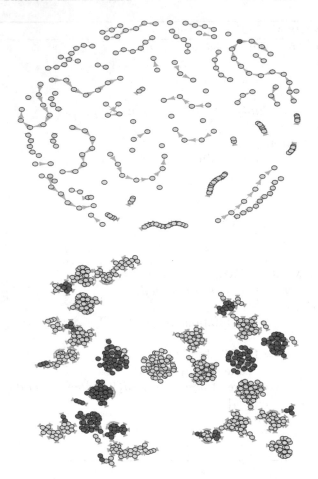

Fig. 1. Top 2% of local optima which were sampled using *Markov-Chain* LON Construction (top) and top 0.05% sampled during *Hybrid* LON Construction (bottom). Pseudo-global optima (i.e. the best in that particular sample for the purposes of this visualisation) are shown in red; all other local optima are grey. For the *Markov-Chain* LON, the highest fitness is 1.707677. For the *Hybrid* LON, it is 1.707826. (Color figure online)

On the higher LON, the sequence is perturbation ⟶ local search; on the bottom, recombination ⟶ probabilistic random mutation ⟶ local search. Nodes with the highest fitness in that sample are red; all other local optima are grey.

Examining the Figure, *Markov-Chain* is the sparser network of the two. There are neat sequences of local optima and nodes typically have one incoming edge and one outgoing edge. The sequences are separate, in that they do not have bridges connecting them. The highest-fitness node is located within a sequence. The visual analysis of this implies that this fitness would only be reached if the search arrived by happenstance on that particular sequence of local optima.

The *Hybrid* LON is denser and instead of linear sequences, clusters of nodes are seen. Some clusters are connected to other clusters and some are isolated. Many distinct solutions with the pseudo-optimal fitness are found by the *Hybrid* (look at the red nodes) and these are found in different clusters. The presence of clusters instead of linear sequences hints that at lower fitness levels (not shown) the clusters would be larger and more opportunity to connect to a pseudo-optimum would be found when comparing to the *Markov-Chain* LON.

4 Experimental Setup

4.1 *Markov-Chain* LON Construction: Details

As stipulated in Sect. 2.1, *Markov-Chain* LON Construction algorithm is an Iterated Local Search (ILS) framework. As such, local search handles intensification and perturbation mechanism contributes diversification. The local search uses a bit-flip operator and best-improvement as a pivot rule. A solution is deemed a local optimum at the end of 100 iterations. Perturbation is 30 bit-flips and improving local optima are always accepted. Deteriorating local optima are accepted 10% of the time. Runs terminate after 1000 iterations. Thirty independent runs are conducted, with each accepted local optimum added as a LON node and each transformation between two local optima added as a LON edge. The parameters are shown in Table 1.

Table 1. *Markov-Chain* LON Construction parameter settings

Parameter	Value
Local iterations	100
Global iterations	1000
Pivot rule	Best
Local search	1 bit-flips
Perturbation	30 bit-flips
Number of runs	30

Iterated Local Search. The ILS process from *Markov-Chain* LON Construction is modified (without any LON logging) and is proposed for optimisation of CSOP. We use it in collecting algorithm performance information. The algorithmic setting remains the same except the best-improvement rule changes to the best of 100.

4.2   Hybrid LON Construction: Details

Hybrid LON Construction is instrumented on top of a competitive GA for the domain [13]. A previous study using statistical inference found that only two GA parameters were significant on this CSOP when solutions are binary-encoded: crossover probability $\phi\prime$ and mutation probability $\phi\prime\prime$ [21]. Our values for those parameters are the ones they recommended ($\phi\prime = 0.614, \phi\prime\prime = 0.198$); the others are from a related study [14] (which used integer encoding for the problem), in the absence of reported values in the binary-encoded study. A random starting population of 76 individuals, all binary strings with $n = 400$, is created. Elitism is implemented for the fittest two individuals; the selection method is linear roulette-wheel (parents are selected with probability proportionate to their fitness ranking); selection pressure is seven; and there are six points of crossover, with the crossover type being uniform. We added local search, making the algorithm memetic. The local search was best-improvement, using single bit-flips, and for 100 iterations. This is applied to the best 10% of individuals at each generation. Those individuals are added as LON nodes, recombined, and the trajectories from parent to child are taken as LON edges. After 100 generations, pairwise combinations of LON nodes are recursively recombined 10 times: offspring from the first recombination are subject to local search and then become the parents for the next. Nodes and edges are added to the LON during this process.

Memetic Search. A variant of the MS framework described in the previous Section (without any LON logging) is also proposed here for optimisation on the CSOP. In our study we use it to collect algorithm performance information. The percentage of individuals locally optimised at each generation becomes 5%; the best-improvement local search becomes first-improvement; and the local search operator becomes ten bit-flips instead of one. All parameter settings for the GA component remain the same. 100 generations are allowed.

4.3   Unknown Global Optimum

For this problem the global optimum is not known. In previous LON research, there has always been a known optimum fitness. To simulate this for our problem, we conduct several runs of the MS and ILS and take the highest obtained fitness across all runs to be the pseudo-optimal fitness. This value is 1.71.

5   Results

5.1   The Hybrid LON

The *Hybrid* LON Construction network has 124,497 nodes and 1,264,500 edges, giving an edge-to-node ratio of 10:1. The average fitness is 0.909698, which at above zero is a feasible solution and is around 53% of the pseudo-optimal fitness stipulated in Sect. 4.3. The maximum fitness is 1.707826, which is within 0.001% of the pseudo-optimal fitness. The minimum fitness (of a local optimum in the sample) is −106.717. There are 217 different solutions with the pseudo-optimal fitness. The vast majority—93.6%—of the local optima are feasible solutions.

Around 14.5% of edges in the LON represent no fitness change; 43.6% are improving fitness; and 41.8% have deteriorating fitness.

The *assortativity* coefficient of a network is the Pearson's correlation for the degrees of connected nodes. In the *Hybrid*-constructed LON, it stands at 0.794687. This implies that it is likely for a node to connect to nodes which have similar degree.

The median degree for a node in the LON is 10; the mean is 20.31; the 75% quantile is 16; and the maximum is very large at 179,154. Most nodes have relatively low degree (≤ 16) and only 0.1% of nodes have degree ≥ 241. The presence of a single node with excessively high degree (179154) hints at a hub-and-spoke system being present in a section of the LON.

5.2   The Markov-Chain LON

The *Markov-Chain* LON Construction models an adaptive walk through the LOL. There are 11,393 nodes and 209,489 edges in the sample, for an edge-to-node ratio of approximately 20:1. The average sampled fitness is 0.638913—around 37% of the pseudo-optimal fitness. This is noticeably lower than the average fitness in the *Hybrid* LON. The maximum fitness is 1.707677, which is lower than both the pseudo-optimal fitness and the maximum fitness in the *Hybrid* LON but is still within 0.002% of the pseudo-optimal value. The minimum is −61.6745, which is approximately twice as fit as the lowest in the *Hybrid* LON. This makes sense given the unguided nature of selection in the *Hybrid* algorithm compared to the guided walk of the *Markov-Chain* process.

In the LON, around 64% of edges are deteriorating (that is, they orient towards a worse fitness); 26% are improving; and around 9% direct towards equal fitness. The majority are deteriorating even though deteriorating moves are only accepted 10% of the time. This fact hints at the scarcity of improving moves on the local optima level manifesting under these operators. Let us compare the percentages with those present in the *Hybrid* LON (43.6% improving and 41.8% deteriorating, as we recall); a judicious conclusion is that the recombination $\longrightarrow$ local search sequence of the *Hybrid* algorithm has more *evolvability* potential on the LOL than the perturbation $\longrightarrow$ local search sequence of the *Markov-Chain* algorithm.

The assortativity coefficient is 0.996704, which stipulates that nodes are highly likely to be connected to nodes which have the same degree as them. This is evidence against the presence of a 'hub-and-spoke' network structure in this LON because that phenomenon is defined by heterogeneous degree distribution. The median degree in the LON is 34; the mean is close by at around 37; the 75% quantile is 52; and the maximum degree is 526. Only 0.01% of nodes have degree ≥ 128. The range of values in the degree distribution is much less extreme than was present in the *Hybrid* LON.

5.3   A Study of Feasibility in LONs

The existence of infeasible solutions in CSOP fitness landscapes brings new possibilities for the features calculated from the LONs. One consideration is the proportion of LON nodes which are infeasible. In the *Hybrid* LON Construction object, 93.6% of nodes are feasible (meaning they have fitness above 0.0000); the *Markov-Chain* LON Construction object has 86.2% feasible nodes. This implies that the former more heavily exploits feasible regions at the level of local optima. Another detail that can be studied for the LONs is the notion of *feasibility gradient*. This is the change in fitness feasibility that a LON edge encodes. An edge could be oriented from an infeasible local optimum towards a feasible local optimum, which is a desirable situation. The distribution of feasibility gradients in the LON therefore captures the ability of the LON algorithm to escape infeasible regions.

Table 2. Percentage of *Markov-Chain* LON edges in terms of feasibility gradient

Orientation	Percentage
infeasible ⟶ feasible	≈77%
infeasible ⟶ infeasible	≈14%
feasible ⟶ feasible	≈5%
feasible ⟶ infeasible	≈4%

Table 3. Percentage of *Hybrid* LON edges in terms of feasibility gradient

Parameter	Value
feasible ⟶ feasible	≈84.8%
infeasible ⟶ feasible	≈7.7%
infeasible ⟶ infeasible	≈4.1%
feasible ⟶ infeasible	≈3.4%

In Table 2 are indications of the feasibility gradients (in percentage terms) of the edges of the sampled *Markov-Chain* LON. These must be viewed with the consideration that the algorithm used to construct the edges always accepts improving local optima, but also accepts deteriorating local optima 10% of the time. Encouragingly, the large majority (77%) of edges orient from infeasible to feasible local optima. That implies the operator sequence often succeeds in traversing portals out of infeasible regions. Transformations from feasible ⟶ feasible are much fewer at approximately 5% of total edges. This perhaps implies that the operator sequence is not great at exploiting within the feasible regions in the search space.

Table 3 shows the feasibility gradient percentages seen in the *Hybrid* LON. Here a vast majority (84.8%) of the orientations are from feasible ⟶ feasible. This hints the operator sequence is proficient at intensification within promising areas in the search space. The percentage of directions from infeasible ⟶ feasible is small, which could also be important—maybe the algorithm struggles to escape infeasible areas. It could be, however, that this small percentage is born from the fact that the number of infeasible nodes in the network is low. A surprisingly low percentage (3.4%) lead from feasible ⟶ infeasible solutions. This is interesting, because there is no acceptance condition for nodes during the construction. It seems recombining already-fit solutions before refining the offspring with local search results in fit solutions.

5.4   Newly Proposed Optimisation Algorithms

As described in Sects. 4.1 and 4.2 we propose and use modified versions of our ILS and MS frameworks to conduct optimisation on the CSOP and compare algorithm performances, alongside the GA from the literature [21]. Tables 4 and 5 summarise distributions for the obtained fitness (averaged over 100 runs) for each algorithm variant. Table 4 displays algorithm results when not enforcing a fitness evaluation budget; Table 5 shows results from the versions which were budgeted 50 000 fitness evaluations. Each row is an algorithm variant. Indications of the variant are found in the *algorithm* and *seeded* columns. By *seeded*, we mean that for these runs a specific solution was seeded into the algorithms. The solution was not chosen due to good fitness (in fact, the fitness is infeasible and heavily-penalised) but rather to provide the same solution across algorithms for 100 runs. In the case of GA and MA, this solution was one individual in the starting populations; for the ILS, it served as the starting solution. As asserted in Sect. 4.3, we assume the pseudo-optimal fitness value of 1.71 for the purposes of this study.

Table 4. Averaged obtained fitness over 100 runs of the algorithms. In the case of the EAs, this is the best fitness in the population. No computational budget is specified.

Algorithm	Seeded	Minimum	1st quantile	Median	Mean	3rd quantile	Maximum
ILS	No	−25.43	−7.39	−2.09	−4.49	−0.48	**1.66**
GA	No	−71 556.39	−97.64	−32.15	−3738.60	−11.67	**0.40**
MS	No	**1.46**	**1.70**	**1.71**	**1.70**	**1.71**	**1.71**
ILS	Yes	−62.54	−6.61	−1.87	−4.86	−0.31	**1.67**
GA	Yes	−137 600.69	−212.50	−32.79	−4084.30	−7.55	−0.19
MS	Yes	**1.45**	**1.70**	**1.71**	**1.68**	**1.71**	**1.71**

Across budgeted and non-budgeted runs, seeded and unseeded runs, the Genetic Algorithm (GA) is definitely the least competent at obtaining fit (or even feasible) solutions. This can be seen by, for example, comparing the median

Table 5. Averaged obtained fitness over 100 runs of the algorithms. In the case of the EAs, this is the best fitness in the population. The computational budget for each run is 50 000 function evaluations.

Algorithm	Seeded	Minimum	1st quantile	Median	Mean	3rd quantile	Maximum
ILS	No	−55.89	−10.22	−4.26	−7.38	−0.64	**1.66**
GA	No	−235 584.20	−2538.16	−75.86	−12 476.17	−32.63	−4.07
MS	No	**1.15**	**1.46**	**1.66**	**1.59**	**1.68**	**1.70**
ILS	Yes	−46.84	−6.25	−1.96	−4.18	**0.15**	**1.44**
GA	Yes	−271 039.28	−4572.31	−63.20	−10 765.08	−27.74	**1.19**
MS	Yes	**1.31**	**1.45**	**1.67**	**1.60**	**1.68**	**1.70**

of the GA rows with the median of the MS or ILS rows in either of Table 4 or Table 5. Sometimes the difference in fitness is several orders of magnitude—see for example the algorithm comparison available *minimum* columns.

The MS performs by far the best of the three algorithms. In all cases (budgeted and unbudgeted, seeded and unseeded), 100% of the runs ended with a feasible fitness found in the population. That is shown in the third and sixth rows of both tables. The best ILS runs (the *maximum* rows) ended with a desirable fitness although the majority ended with infeasible fitness. It is of note, however, that the distributions comprise fitness values which are 'almost' feasible in many cases. The deduction seems to be that the success (or lack thereof) of ILS on this CSOP depends on the starting position. The 'almost' feasible fitness values, we argue, are the dead-ends of sub-optimal *massif centrals* or *funnels*.

The success of the MS tells us that using all of: recombination, random mutation, and guided local search together works in harmony with this configuration space to guide the search to promising feasible regions. The genetic algorithm's vast range of obtained fitnesses suggests a lack of reliability. Sometimes a feasible solution may be found (a previous paper found that it was around 5% of runs given 100 generations [14]) but other times a population filled with individuals of severely penalised fitness may be obtained. Contrarily, the MA appears to be rather uniformly consistent: all of the total 400 runs ended with a feasible fitness, and this was also always ≥1.15.

The consistency of the ILS lies somewhere between the performances of the GA and the MS: although often the end fitness is infeasible, the range of values in the distributions is tight compared to the GA and are usually between −10 and +1.66. ILS solutions could be seeded to a different highly-exploitative algorithm to finish the job.

6   Conclusions

This work has pursued modelling a problem from healthcare with LONs. Doing so brought the added complications of infeasible regions in the fitness landscapes, which is new for LON analysis. Two algorithms were offered for the purpose of constructing LONs for CSOP: *Markov-Chain* LON Construction and *Hybrid* LON Construction. An examination of the feasibility gradients within the LONs revealed that an ILS framework (i.e. the *Markov-Chain* algorithm) may be better at finding portals out of infeasible regions but lack in local optima-level exploitative power when in a promising region. MS (i.e. the *Hybrid* algorithm) appears very proficient in exploitation within feasible areas but did not boast many escapes from infeasible areas. This provides insight into how CSOP interacts with sequences of search operators. We showed that our MS and our ILS outperformed the GA from the literature, even when mandating an equal fitness function budget for the algorithms. We stipulate that the selection process of the GA does not have sufficient exploitative power (perhaps due to the small population size) and this can be brought by adding local search. Importantly, the best results are obtained using the recombination and random mutation of the GA *together with* a guided local search. It follows that the former bring innovation and diversification, while the latter brings intensification and facilitates propagation of good genes.

References

1. Ochoa, G., Tomassini, M., Vérel, S., Darabos, C.: A study of NK landscapes' basins and local optima networks. In: Proceedings of the 10th Annual Conference on Genetic and Evolutionary Computation, pp. 555–562. ACM (2008)
2. Verel, S., Ochoa, G., Tomassini, M.: The connectivity of NK landscapes' basins: a network analysis. arXiv preprint arXiv:0810.3492 (2008)
3. Herrmann, S., Ochoa, G., Rothlauf, F.: Communities of local optima as funnels in fitness landscapes. In: Proceedings of the 2016 on Genetic and Evolutionary Computation Conference, pp. 325–331. ACM (2016)
4. Daolio, F., Tomassini, M., Vérel, S., Ochoa, G.: Communities of minima in local optima networks of combinatorial spaces. Phys. A **390**(9), 1684–1694 (2011)
5. Iclanzan, D., Daolio, F., Tomassini, M.: Data-driven local optima network characterization of QAPLIB instances. In: Proceedings of the 2014 Annual Conference on Genetic and Evolutionary Computation, pp. 453–460. ACM (2014)
6. Verel, S., Daolio, F., Ochoa, G., Tomassini, M.: Sampling local optima networks of large combinatorial search spaces: the QAP case. In: Auger, A., Fonseca, C.M., Lourenço, N., Machado, P., Paquete, L., Whitley, D. (eds.) PPSN 2018. LNCS, vol. 11102, pp. 257–268. Springer, Cham (2018). https://doi.org/10.1007/978-3-319-99259-4_21
7. Ochoa, G., Veerapen, N., Whitley, D., Burke, E.K.: The multi-funnel structure of TSP fitness landscapes: a visual exploration. In: Bonnevay, S., Legrand, P., Monmarché, N., Lutton, E., Schoenauer, M. (eds.) EA 2015. LNCS, vol. 9554, pp. 1–13. Springer, Cham (2016). https://doi.org/10.1007/978-3-319-31471-6_1

8. Veerapen, N., Ochoa, G., Tinós, R., Whitley, D.: Tunnelling crossover networks for the asymmetric TSP. In: Handl, J., Hart, E., Lewis, P.R., López-Ibáñez, M., Ochoa, G., Paechter, B. (eds.) PPSN 2016. LNCS, vol. 9921, pp. 994–1003. Springer, Cham (2016). https://doi.org/10.1007/978-3-319-45823-6_93

9. Ochoa, G., Veerapen, N.: Mapping the global structure of TSP fitness landscapes. J. Heuristics **24**, 265–294 (2018). https://doi.org/10.1007/s10732-017-9334-0

10. Simoncini, D., Barbe, S., Schiex, T., Verel, S.: Fitness landscape analysis around the optimum in computational protein design. In: Proceedings of the Genetic and Evolutionary Computation Conference, pp. 355–362. ACM (2018)

11. Mostert, W., Malan, K.M., Ochoa, G., Engelbrecht, A.P.: Insights into the feature selection problem using local optima networks. In: Liefooghe, A., Paquete, L. (eds.) EvoCOP 2019. LNCS, vol. 11452, pp. 147–162. Springer, Cham (2019). https://doi.org/10.1007/978-3-030-16711-0_10

12. McCall, J., Petrovski, A.: A decision support system for cancer chemotherapy using genetic algorithms. In: Proceedings of the International Conference on Computational Intelligence for Modeling, Control and Automation, pp. 65–70 (1999)

13. Petrovski, A.: An application of genetic algorithms to chemotherapy treatment (1998)

14. Petrovski, A., Brownlee, A., McCall, J.: Statistical optimisation and tuning of GA factors. In: 2005 IEEE Congress on Evolutionary Computation, vol. 1, pp. 758–764. IEEE (2005)

15. Petrovski, A., Shakya, S., McCall, J.: Optimising cancer chemotherapy using an estimation of distribution algorithm and genetic algorithms. In: Proceedings of the 8th Annual Conference on Genetic and Evolutionary Computation, pp. 413–418. ACM (2006)

16. McCall, J., Petrovski, A., Shakya, S.: Evolutionary algorithms for cancer chemotherapy optimization. In: Computational Intelligence in Bioinformatics, vol. 7, p. 265 (2007)

17. Brownlee, A.E., Pelikan, M., McCall, J.A., Petrovski, A.: An application of a multivariate estimation of distribution algorithm to cancer chemotherapy. In: Proceedings of the 10th Annual Conference on Genetic and Evolutionary Computation, pp. 463–464. ACM (2008)

18. Wheldon, T.E.: Mathematical Models in Cancer Research. Taylor & Francis, Abingdon (1988)

19. Tse, S.M., Liang, Y., Leung, K.S., Lee, K.H., Mok, T.S.K.: A memetic algorithm for multiple-drug cancer chemotherapy schedule optimization. IEEE Trans. Syst. Man Cybern. Part B (Cybern.) **37**(1), 84–91 (2007)

20. Barbour, R., Corne, D., McCall, J.: Accelerated optimisation of chemotherapy dose schedules using fitness inheritance. In: IEEE Congress on Evolutionary Computation, pp. 1–8. IEEE (2010)

21. Petrovski, A., Wilson, A., Mccall, J., et al.: Statistical identification and optimisation of significant GA factors. In: Proceedings of the 5th Joint Conference on Information Sciences, Atlantic City, USA, vol. 1, pp. 1027–1030 (2000)

22. Petrovski, A., McCall, J.: Multi-objective optimisation of cancer chemotherapy using evolutionary algorithms. In: Zitzler, E., Thiele, L., Deb, K., Coello Coello, C.A., Corne, D. (eds.) EMO 2001. LNCS, vol. 1993, pp. 531–545. Springer, Heidelberg (2001). https://doi.org/10.1007/3-540-44719-9_37

23. Villasana, M., Ochoa, G.: Heuristic design of cancer chemotherapies. IEEE Trans. Evol. Comput. **8**(6), 513–521 (2004)

24. Agur, Z., Hassin, R., Levy, S.: Optimizing chemotherapy scheduling using local search heuristics. Oper. Res. **54**(5), 829–846 (2006)

25. Ochoa, G., Villasana, M., Burke, E.K.: An evolutionary approach to cancer chemotherapy scheduling. Genet. Program Evolvable Mach. **8**(4), 301–318 (2007). https://doi.org/10.1007/s10710-007-9041-y
26. Ochoa, G., Veerapen, N.: Deconstructing the big valley search space hypothesis. In: Chicano, F., Hu, B., García-Sánchez, P. (eds.) EvoCOP 2016. LNCS, vol. 9595, pp. 58–73. Springer, Cham (2016). https://doi.org/10.1007/978-3-319-30698-8_5
27. Ochoa, G., Herrmann, S.: Perturbation strength and the global structure of QAP fitness landscapes. In: Auger, A., Fonseca, C.M., Lourenço, N., Machado, P., Paquete, L., Whitley, D. (eds.) PPSN 2018. LNCS, vol. 11102, pp. 245–256. Springer, Cham (2018). https://doi.org/10.1007/978-3-319-99259-4_20

Genetic Programming with Adaptive Search Based on the Frequency of Features for Dynamic Flexible Job Shop Scheduling

Fangfang Zhang[1](✉) ⓘ, Yi Mei[1] ⓘ, Su Nguyen[2] ⓘ, and Mengjie Zhang[1] ⓘ

[1] School of Engineering and Computer Science, Victoria University of Wellington,
PO BOX 600, Wellington 6140, New Zealand
{fangfang.zhang,yi.mei,mengjie.zhang}@ecs.vuw.ac.nz
[2] Centre for Data Analytics and Cognition, La Trobe University,
Melbourne, VIC 3086, Australia
P.Nguyen4@latrobe.edu.au

Abstract. Dynamic flexible job shop scheduling (DFJSS) is a very valuable practical application problem that can be applied in many fields such as cloud computing and manufacturing. In DFJSS, machine assignment and operation sequencing decisions need to be made simultaneously in dynamic environments with unpredicted events such as new job arrivals. Scheduling heuristic is an ideal candidate for solving the DFJSS problem due to its efficiency and simplicity. Genetic programming (GP) has been successfully applied to evolve scheduling heuristics for job shop scheduling automatically. However, GP has a huge search space, and the traditional search algorithms do not utilise effectively the information obtained from the evolutionary process. This paper proposes a new method to make better use of the information during the evolutionary process of GP to further enhance the ability of GP. To be specific, this paper proposes two adaptive search strategies based on the frequency of features in promising individuals to guide GP to evolve effective rules. This paper examines the proposed algorithm on six different DFJSS scenarios. The results show that the proposed GP with adaptive search can converge faster and achieve significantly better performance than the GP without adaptive search in most scenarios while no worse in all other scenarios without increasing the computational cost.

Keywords: Adaptive search · Scheduling heuristic · Dynamic flexible job shop scheduling · Genetic programming

1 Introduction

Job shop scheduling (JSS) [1] is an important combinational optimisation problem, which has essential roles in all walks of life such as manufacturing [2,3] and cloud computing [4]. The task in JSS is to process a number of jobs by a set

© Springer Nature Switzerland AG 2020
L. Paquete and C. Zarges (Eds.): EvoCOP 2020, LNCS 12102, pp. 214–230, 2020.
https://doi.org/10.1007/978-3-030-43680-3_14

of machines. Each job has a sequence of operations. The goal is to optimise the machine resources to achieve the objectives such as minimising the max-flowtime. Flexible JSS (FJSS) [5] is a variant of JSS which is better to reflect requirements in real-world applications. In FJSS, an operation can be processed on a set of machines. It indicates that two decisions need to be made simultaneously. One is *machine assignment* (i.e. assign an operation to a particular machine), and the other is *operation sequencing* (i.e. choose an operation as the next operation to be processed by an idle machine). In addition, many practical scheduling problems are dynamically changing over time, for example, due to new job arrivals [6,7]. Dynamic FJSS (DFJSS) is to consider FJSS under dynamic environments.

Scheduling heuristics such as dispatching rules [8] are widely used to handle such kinds of dynamic problems. A scheduling heuristic is a heuristic that works like a priority function to evaluate the priorities of operations and machines. To be specific, in DFJSS, a machine that has the highest priority value based on the *routing rule* (i.e. routing scheduling heuristic) will be assigned a job. An operation with the highest priority value based on the *sequencing rule* (i.e. sequencing scheduling heuristic) will be chosen as the next operation to be processed. There are some rules such as SPT (i.e. shortest processing time) and WIQ (i.e. the workload in the queue of a machine) which have been identified as effective rules for JSS. However, they are manually designed by experts, which is time-consuming and not always available. In practical, it is hard to manually design effective rules due to the complexity of the job shop environments.

Genetic programming (GP) [9], as a hyper-heuristic (GPHH) method, has been successfully applied to automatically evolve scheduling heuristic for JSS [10,11]. As a population-based algorithm, GP tries to improve the scheduling heuristics (i.e. individuals) generation by generation. In traditional GP, features are randomly chosen to build subtrees for mutation and generate individuals. However, the importance of features can be different. Such a way that chooses all the features randomly cannot fully play its role because the importance of the features is ignored. The challenge is that the search space of GP is huge (i.e. the individual can be a big tree), and the traditional search might not be enough. This paper proposes the adaptive search to guide GP to the more promising region by utilising the information during the evolutionary process. The proposed algorithm aims to guide the behaviour of GP over time adaptively.

The key to the success of GP is that it can automatically detect important features and optimise the structure of individuals guided by the fitness function. From an evolutionary perspective, the individuals themselves, especially good individuals, contain useful information which can be further utilised to improve evolutionary efficiency. An advantage is that information generated during the evolutionary process can be easily used without putting more extra effort to get the information. In this paper, the frequency of features based on the individuals that have good performance will be further used to guide GP to find more effective rules for DFJSS adaptively. To this end, two adaptive search strategies which can be realised by mutation and re-initialisation will be proposed.

The overall goal of this paper is to develop effective adaptive search strategies with the frequency of features to guide GP to find effective scheduling heuristics for DFJSS efficiently. The proposed algorithms are expected to speed up the convergence of GP and find effective rules without additional computing requirement. In particular, this paper has the following research objectives:

- Develop adaptive search strategies with the frequency of features in promising individuals to guide GP towards the more promising areas.
- Verify the effectiveness and efficiency of the proposed GP algorithm with the adaptive strategy by comparing its performance and convergence speed with the baseline GP.
- Analyse how the adaptive search affects the evolutionary process of GP.

2   Background

2.1   Dynamic Flexible Job Shop Scheduling

In FJSS problem, n jobs $J = \{J_1, J_2, \ldots, J_n\}$ need to be processed by m machines $M = \{M_1, M_2, \ldots, M_m\}$. Each job J_j has an arrival time $at(J_i)$ and a sequence of operations $O_j = (O_{j1}, O_{j2}, \ldots, O_{ji})$. Each operation O_{ji} can only be processed by one of its optional machines $\pi(O_{ji})$ and its processing time $\delta(O_{ji})$ depends on the machine that processes it. It indicates that there are two decisions which are routing decision and sequencing decision in FJSS. In DFJSS, not only two decisions need to be made simultaneously, but also the dynamic events are necessary to be taken into account when making schedules. This paper focuses on one dynamic event (i.e. continuously arriving new jobs). That is, the information of a job is unknown until its arrival time.

2.2   Genetic Programming Hyper-heuristic for DFJSS

A hyper-heuristic [12] is a heuristic search method that seeks to select or generate heuristics to efficiently solve hard problems. The unique characteristic is that hyper-heuristic works on heuristic search space instead of solution search space.

GP, as a hyper-heuristic method [13], has been successfully applied to more informative scheduling heuristics for combinational optimisation problems such as packing [14,15], timetabling [16,17] and JSS [18–21]. Scheduling heuristics, including routing and sequencing rules, are needed in DFJSS in our research. To follow the sequence constraint of operations of a job, this paper only starts to allocate an operation when it becomes a *ready operation*. There are two sources of ready operations. One is the first operation of a job. The second is the operation that its proceeding operation is just finished. Once an operation becomes a ready operation (*routing decision point*), it will be allocated to the machine by the routing rule. When a machine becomes idle, and its queue is not empty (*sequencing decision point*), the sequencing rule will be triggered to choose the next operation to be processed.

GP has shown its superiority in DFJSS [18,19]. However, most of the existing works follow the traditional way of the evolutionary process of GP, which may not enough due to its large search space. To this end, this paper introduces the adaptive search to help GP evolve more effective scheduling heuristics (i.e. routing rule and sequencing rule) for DFJSS.

3   The Proposed GP with Adaptive Search

In this paper, the adaptive search aims to guide the behaviour of the GP algorithm over time by utilising the information generated during the evolutionary process of GP. It is expected to speed up the convergence of GP and evolve effective rules. It is not trivial to answer "when", "how", and "where" to apply the adaptive search. These three research questions are investigated as follows.

Question 1: When to use the adaptive search?

This paper uses the adaptive strategy at every generation.

Question 2: How to use the adaptive search?

In this paper, the frequency of features is the number of occurrences of features. The number of occurrences of features based on the entire top ten individuals (i.e. roughly 1% of the population size) will be utilised to guide the behaviour of GP to improve its convergence speed and find more promising rules for DFJSS, since top ten individuals have much better fitnesses than others. Based on the number of occurrences of features, the probability of each feature is calculated. The larger the number of the occurrence, the higher the probability that the feature is given. When generating new individuals and subtrees for mutation, the features will be selected based on their probability. The higher the probability, the easier the feature is to be selected for building new trees. The pseudo-code of calculating the probabilities of features are shown in Algorithm 1.

Algorithm 1. Pseudo-code of calculating the probabilities of features

---

Input   : Top ten individuals
Output: The probabilities of features *probabilities*
1: *probabilities* ← *null*
2: **for** $i = 1$ to $|featureSize|$ **do**
3: │   *occurrence$_i$*: count the number of occurrence of a feature in the *top ten individuals*
4: **end**
5: *sumOccurrences*: sum up the occurrences of all features
6: **for** $i = 1$ to $|featureSize|$ **do**
7: │ $probability_i = \frac{occurrence_i}{sumOccurrences}$
8: **end**
9: *probabilities* ← *probability$_i$*
10: **return** *probabilities*

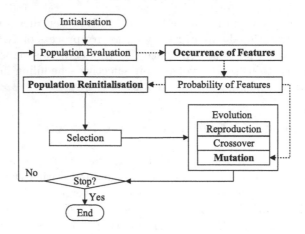

Fig. 1. The flowchart of the proposed GP with the adaptive search.

Question 3: Where to use the adaptive search?

During the evolutionary process of GP, there are two occasions using selected features. One is when building subtree for mutation. The other is when generating new individuals. To guide the behaviour of GP as much as possible, it is straightforward to simply apply the adaptive search by mutation. Another adaptive search strategy related to re-initialisation is also proposed in this paper.

Mutation. As a genetic operator, mutation aims to maintain the diversity of the population by replacing one subtree with a randomly generated subtree. The new individual produced by mutation can be very bad (i.e. too random). If the mutation direction can be guided to some extent, it may enhance the effectiveness of mutation. The subtree that builds with the informed features has such a role because it has a high chance to bring more useful building blocks.

Re-initialisation. The quality of individuals in the population can be different. Some individuals have good performance and have a higher chance to be selected as parents to generate offspring. However, there still has a number of individuals that will not contribute too much to the next generation due to their lower performance. These kinds of individuals are useless to some extent. This paper proposes to use a re-initialisation strategy to generate some useful individuals to the population to replace them at each generation. The reinitialised new individuals generated based on the frequency of features are structurally different and have reasonably good fitness.

It is not trivial to decide which individuals to remove from the population. Removing too many individuals takes the risk to lose the quality of the population. Removing very few individuals may not work. This paper uses simulation information to decide which individuals will be removed. If an individual leads to a very long queue of a machine during the simulation, it will be replaced by randomly generated individuals based on the frequency of features. This is because the evolved best rules will not assign a machine lots of operations from our preliminary investigation. In this way, the current population is more likely to have more promising individuals, and thus more capable of generating better offspring for the next generations.

Figure 1 depicts the general outline of the proposed GP with the adaptive search. GP starts to initialise the population randomly, and then evaluate the individuals in the population. The individuals in the new generation (i.e. off-spring) are generated in the evolution stage along with the selection. For the proposed algorithm, there are three main differences compared with standard GP. The first one is that the frequency of features are counted based on the entire top ten individuals after evaluating the individuals. The top ten individuals are better than others obviously, which are good for measuring the frequency of features from our preliminary work. This information is converted into a probability for each feature. The second one is that the re-initialisation strategy is applied to import new potential good individuals into the population by generating new individuals based on the frequency of features. The last one is that the frequency of feature information is utilised to guide the mutation direction. The adaptive strategies are applied over time.

4   Experiment Design

4.1   Simulation Model

Simulation is a common method to investigate complex real-world problems [22]. This paper assumes there are 5000 jobs that need to be processed by ten machines. The importance of jobs might be different, which are indicated by weights. The weights of 20%, 60%, and 20% of jobs are set as one, two and four, respectively. The number of operations of each job varies by a uniform discrete distribution between one and ten. The processing time of each operation is set by uniform discrete distribution with the range [1]. The number of candidate machines for an operation follows a uniform discrete distribution between one and ten.

In each problem instance, jobs arrive stochastically according to a Poisson process with rate λ. To improve the generalisation ability of the evolved rules for DFJSS problems, the seeds used to generate the jobs are rotated at each generation. In addition, to make sure the accuracy of the collected data, a warm-up period of 1000 jobs is used. If any machine in the system has more than 100 operations, the simulation will be stopped, and the current evaluating individual is replaced by a new individual based on re-initialisation strategy.

4.2   Parameter Settings

In our experiment, the terminal and function set are shown in Table 1, following the setting in [23]. The "/" operator is protected division, returning one if divided by zero. The other parameter settings of GP are shown in Table 2.

Table 1. The terminal and function sets.

	Series terminals	Series description
Machine-related	NIQ	The number of operations in the queue
	WIQ	Current work in the queue
	MWT	Waiting time of a machine
Operation-related	PT	Processing time of an operation
	NPT	Median processing time for next operation
	OWT	The waiting time of an operation
Job-related	WKR	The median amount of work remaining of a job
	NOR	The number of operations remaining of a job
	W	Weight of a job
	TIS	Time in system
Functions	$+, -, *, /, max, min$	As usual meaning

Table 2. The parameter setting of GP.

Parameter	Value
Number of subpopulations	2
Subpopulation size	512
Method for initialising population	Ramped-half-and-half
Initial minimum/maximum depth	2/6
Maximal depth of programs	8
The number of elites	10
Crossover/mutation/reproduction rate	80%/15%/5%
Parent selection	Tournament selection with size 7
Number of generations	51
Terminal/non-terminal selection rate	10%/90%

4.3   Comparison Design

Four algorithms are taken into the comparison in this paper. The cooperative coevolution genetic programming (CCGP) [6] which can be used to evolve routing rule and sequencing rule simultaneously, is selected as the baseline algorithm. Our proposed algorithm, which incorporates with adaptive strategy by mutation, is named as MUGP (i.e. generate subtree based on the frequency of features for mutation). The algorithm that incorporates re-initialisation strategy (i.e. reinitialise some useful individuals based on the machine information during the simulation) is named as IMGP. The proposed algorithm, which incorporated by both mutation and re-initialisation, is named as IM^{2}GP. MUGP, IMGP and IM^{2}GP are compared with CCGP, respectively.

To verify their effectiveness, the proposed algorithms are tested on *six different scenarios*. The scenarios consist of three objectives (e.g. max flowtime, mean

Table 3. The mean (standard deviation) of the objective value of CCGP, MUGP, IMGP, and IM^{2}GP over 50 independent runs for six dynamic flexible scenarios.

Scenario	CCGP	MUGP	IMGP	IM^{2}GP
<Fmax, 0.85>	1212.05(34.68)	1219.73(29.41)	1208.64(30.78)	1219.61(40.52)
<Fmax, 0.95>	1941.98(29.93)	1946.65(47.62)	1934.25(25.90)	1942.72(31.19)
<Fmean, 0.85>	385.95(3.22)	384.79(1.39)	384.89(2.42)(−)	384.57(1.46)(−)
<Fmean, 0.95>	551.18(5.78)	549.66(2.90)	551.20(4.70)	549.51(4.21)(−)
<WFmean, 0.85>	831.41(6.08)	829.43(5.31)(−)	831.20(6.71)	829.26(4.14)(−)
<WFmean, 0.95>	1111.01(12.02)	1107.59(9.57)	1105.44(6.71)(−)	1105.70(6.95)(−)

flowtime, and mean weighted flowtime) and two utilisation levels (e.g. 0.85 and 0.95). For the sake of convenience, Fmax, Fmean, and WFmean are used to indicate max flowtime, mean flowtime, and mean weighted flowtime, respectively. The objective functions are shown as follows.

- Minimising Fmax $= max\{C_1, C_i, \ldots, C_n\}$
- Minimising Fmean $= \frac{\sum_{i=1}^{n}\{C_i - r_i\}}{n}$
- Minimising WFmean $= \frac{\sum_{i=1}^{n} w_i * \{C_i - r_i\}}{n}$

where C_i is the completion time of job J_i, r_i is the release time of J_i, and w_i is the weight of J_i.

Note that the evolved best rule at each generation is tested on 50 different test instances, and the mean objective value is reported as the objective value of this best rule. This aims to guarantee the accuracy of measuring the performance.

5 Results and Discussions

Fifty independent runs are conducted for the comparison. Wilcoxon rank-sum test with a significance level of 0.05 is used to verify the performance of proposed algorithms. In the following results, "−" and "+" indicate the corresponding result is significantly better or worse than its counterpart. If there is no mark there, that means they have similar performance.

5.1 Performance of Evolved Rules

Table 3 shows the mean and standard deviation of the objective value of CCGP, MUGP, IMGP, and IM^{2}GP over 50 independent runs for six dynamic flexible scenarios. The performance of MUGP is significantly better than CCGP in one scenario (e.g. <WFmean, 0.85>. It indicates that the proposed adaptive strategy with the mutation has the potential to take advantage of the information of the frequency of features. However, it does not work in most scenarios. One possible reason is that the mutation rate is too low (i.e. 0.15) to fully utilise the information. The performance of IMGP is significantly better than that of CCGP in only

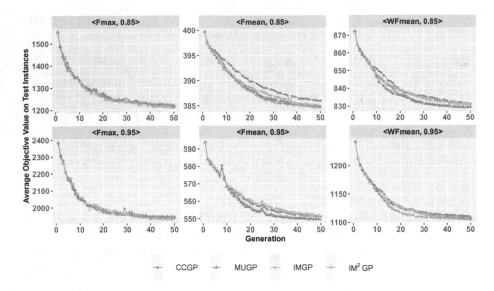

Fig. 2. The convergence curves of CCGP, MUGP, IMGP, and IM^{2}GP over 50 independent runs in six scenarios.

two scenarios (e.g. <Fmean, 0.85> and <WFmean, 0.95>). One possible reason is that the reinitialised individuals do not have a big impact on the population. The performance of IM^{2}GP is significantly better than CCGP in four scenarios (e.g. <Fmean, 0.85>, <Fmean, 0.95>, <WFmean, 0.85> and <WFmean, 0.95>). It indicates that the proposed adaptive strategy with mutation and reinitialisation strategies are more promising. For minimising the max-flowtime, in scenario <Fmax, 0.85> and <Fmax, 0.85>, there is no significant difference among the three algorithms.

Figure 2 shows the convergence curves of the average objective value on the test instances of CCGP, MUGP, IMGP, and IM^{2}GP over 50 independent runs. Except for max-flowtime related scenarios, IM^{2}GP can converge faster and achieve better performance than that of CCGP. For minimising max-flowtime, the proposed three algorithms have no obvious advantages. It might be because max-flowtime is more sensitive to the worst case, which is more complex and hard to optimise.

5.2   Unique Feature Analyses

The number of unique features in the rules is one of the indicators of the complexity of evolved rules. The number of unique feature means the least number of elements that is needed to construct the rules. A rule with a smaller number of features is easier to be interpreted [24].

Figures 3 and 4 show the violin plot of the number of unique features of routing and sequencing rules obtained by CCGP, MUGP, IMGP, and IM^{2}GP over 30 independent runs in different scenarios. Violin plots are similar to box

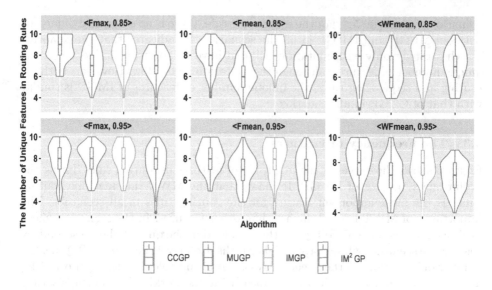

Fig. 3. Violin plot of the number of unique features of *routing rules* obtained by CCGP, MUGP, IMGP, and IM^2GP over 30 independent runs in six scenarios.

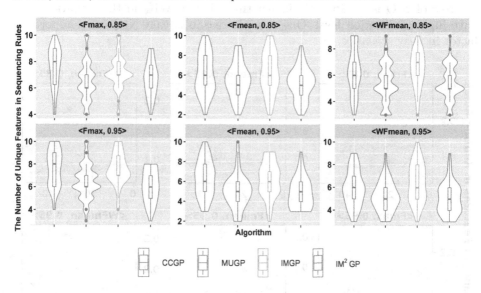

Fig. 4. Violin plot of the number of unique features of *sequencing rules* obtained by CCGP, MUGP, IMGP, and IM^2GP over 30 independent runs in six scenarios.

plots, except that they also show the probability density of the data at different values, usually smoothed by a kernel density estimator. From an overall view, both for the routing rule and sequencing rule, the rules obtained by MUGP and IM^2GP involve a smaller set of features. For the routing rule, there is no statistical difference between MUGP and IM^2GP in most scenarios except the scenario <Fmean, 0.85>. For the sequencing rule, there is no statistical difference between MUGP and IM^2GP in all the scenarios. It indicates that the adaptive search strategy only with mutation still can have a significant influence on the unique number of features, although the mutation rate is low. However, both for routing and sequencing rules, the unique number of features of IMGP is similar with that of CCGP in all scenarios.

5.3 The Frequency of Features

Figure 5 shows the curves of the frequency of features in routing rules during the evolutionary process of IM^2GP. It shows that the MWT (i.e. machine waiting time) is the most important feature for the routing rules in all scenarios. The importance of MWT is much higher than other features. In the scenarios whose utilisation levels are 0.85, WIQ (i.e. the workload in the queue) plays a secondary role. In the scenarios whose have a higher utilisation level (i.e. 0.95), NIQ (i.e. the number of operations in the queue) plays a significant role. Intuitively, both WIQ and NIQ are important indicators for measuring the workload for machines, they might have the same functions, and one might take over another one. However, we do not know how they work in different scenarios. It is interesting to see that the role of NIQ is significantly higher than that of WIQ in the scenarios that have higher utilisation level. One possible reason is that NIQ is an important factor in busy scenarios.

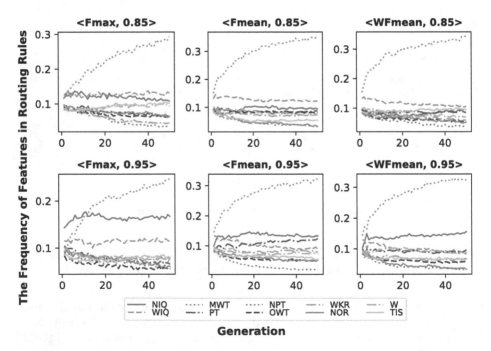

Fig. 5. The curves of the frequency of features in *routing rules* during the evolutionary process of IM^2GP.

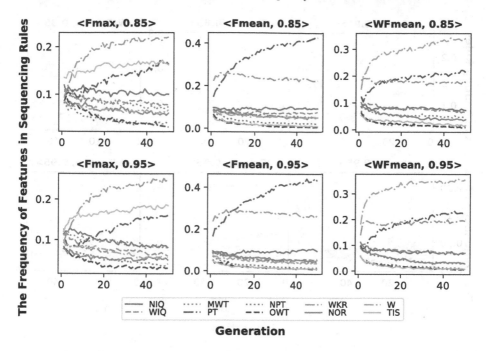

Fig. 6. The curves of the frequency of features in *sequencing rules* during the evolutionary process of IM^2GP.

Figure 6 shows the curves of the frequency of features in sequencing rules during the evolutionary process of IM^2GP. Different from routing rules, there are three features (e.g. WKR, TIS, and PT) play a vital role in minimising max-flowtime. PT and WKR also are two important features in minimising mean-flowtime and weighted mean-flowtime. Except for them, W plays a dominant role in weighted mean-flowtime, which is consistent with our intuition. Besides, W plays its role mainly in sequencing rules instead of routing rules.

It is interesting to see the trend of the feature frequency without adaptive strategies. Figure 7 shows the curves of the frequency of features of the routing rules that evolved by CCGP. Comparing Figs. 5 and 7, both IM^2GP and CCGP can detect important features and use them to build individuals, this is the advantage of GP itself. The difference is that IM^2GP can further enhance this ability. Figure 7 shows that the frequency of feature MWT is much higher than that of CCGP (i.e. the most obvious one) during the evolutionary process (i.e. generation 50). For IM^2GP, in the scenarios with utilisation level 0.85, the frequency of WIQ is higher than other features, which is not that clear for CCGP. Besides, in the scenarios with utilisation level 0.95, the importance of NIQ is easier to be distinguished than that of in CCGP.

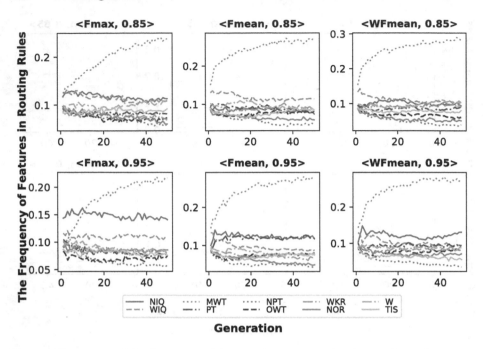

Fig. 7. The curves of the frequency of features in *routing rules* during the evolutionary process of CCGP.

5.4 Reinitialised Individuals

Figure 8 shows the curves of the number of reinitialised routing rules. In all the scenarios, in the beginning, there are a lot of reinitialised individuals in the population. As the number of generations increases, the number of reinitialised routing rules is getting smaller and smaller. After the fifteenth generation, there is no significant change in the number of reinitialised routing rules.

Figure 9 shows the curves of the number of reinitialised sequencing rules. Different from routing rules, the sequencing rules are seldom detected as bad rules. This is in line with our expectations. When evaluating sequencing rules, the best routing rule is used as the collaborator, the probability of a machine that is assigned lots of operations is small. Only when a sequencing rule is quite bad, it might be detected as a bad rule. But for the routing rule, even the best sequencing rule is chosen as the collaborator, there are different routing rules which can lead to a high probability of a machine that is assigned lots of operations.

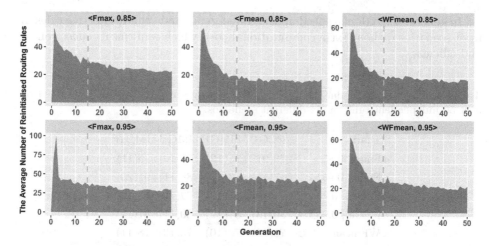

Fig. 8. The curves of the number of *reinitialised routing rules* of IM²GP over 50 independent runs in six different scenarios.

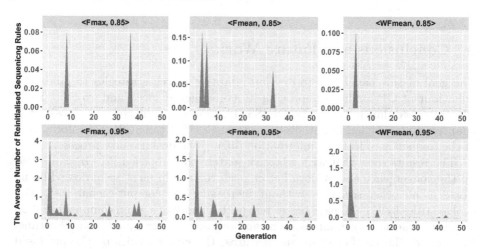

Fig. 9. The curves of the number of *reinitialised sequencing rules* of IM²GP over 50 independent runs in six different scenarios.

5.5   Training Time

Table 4 shows the mean and standard deviation of training time of the four algorithms over 50 independent runs in six scenarios. There is no significant difference between the four algorithms. It means the proposed adaptive search strategies do not need extra computational cost. This verifies the advantages of utilising the information generated during the evolutionary process of GP.

In general, IM^2GP can speed up the convergence and achieve effective rules in most scenarios without extra computational cost, which confirms its effectiveness and efficiency.

Table 4. The mean (standard deviation) of **training time** (in minutes) obtained by over 50 independent runs for six different scenarios.

Scenario	CCGP	MUGP	IMGP	IM^2GP
<Fmax, 0.85>	73(9)	75(12)	75(10)	75(11)
<Fmax, 0.95>	87(15)	83(12)	88(11)	91(23)
<Fmean, 0.85>	71(10)	71(12)	72(11)	73(12)
<Fmean, 0.95>	80(13)	81(12)	80(11)	81(11)
<WFmean, 0.85>	73(13)	73(10)	75(12)	78(11)
<WFmean, 0.95>	82(13)	80(11)	85(10)	87(11)

6   Conclusions and Future Work

The goal of this paper was to develop adaptive search strategies to guide the behaviour of GP for both improving its convergence speed and evolving more effective scheduling heuristics for DFJSS. The goal was achieved by proposing the adaptive mutation and re-initialisation strategies that can utilise the information of the frequency of features information during the evolutionary process.

The results show that with adaptive search, the proposed IM^2GP can speed up the convergence and achieve better performance in most scenarios while no worse in all other scenarios without increasing the computational cost. The evolved rules by IM^2GP have better test performance of a given complex job shop scenario, especially in minimising mean-flowtime and weighted mean-flowtime. In terms of the number of unique features, the evolved rules by the proposed algorithms with adaptive strategies contain fewer features. This can potentially improve the interpretability of the evolved rules because the relationships between features tend to be less complicated. Besides, the proposed algorithms that incorporate the frequency of features information do not need extra computational cost. This shows the benefits of making use of the information during the evolutionary process.

Some interesting directions can be further investigated in the near future. This work already shows the effectiveness to take advantage of the information generated during the evolutionary process. We would like to find more promising ways to detect useful information further to improve its performance.

References

1. Manne, A.S.: On the job-shop scheduling problem. Oper. Res. **8**(2), 219–223 (1960)
2. Geiger, C.D., Uzsoy, R., Aytuğ, H.: Rapid modeling and discovery of priority dispatching rules: an autonomous learning approach. J. Sched. **9**(1), 7–34 (2006). https://doi.org/10.1007/s10951-006-5591-8
3. Tay, J.C., Ho, N.B.: Evolving dispatching rules using genetic programming for solving multi-objective flexible job-shop problems. Comput. Ind. Eng. **54**(3), 453–473 (2008)
4. Nguyen, S.B.S., Zhang, M.: A hybrid discrete particle swarm optimisation method for grid computation scheduling. In: 2014 IEEE Congress on Evolutionary Computation (CEC), pp. 483–490. IEEE (2014)
5. Brucker, P., Schlie, R.: Job-shop scheduling with multi-purpose machines. Computing **45**(4), 369–375 (1990). https://doi.org/10.1007/BF02238804
6. Yska, D., Mei, Y., Zhang, M.: Genetic programming hyper-heuristic with cooperative coevolution for dynamic flexible job shop scheduling. In: Castelli, M., Sekanina, L., Zhang, M., Cagnoni, S., García-Sánchez, P. (eds.) EuroGP 2018. LNCS, vol. 10781, pp. 306–321. Springer, Cham (2018). https://doi.org/10.1007/978-3-319-77553-1_19
7. Zhang, F., Mei, Y., Zhang, M.: Genetic programming with multi-tree representation for dynamic flexible job shop scheduling. In: Mitrovic, T., Xue, B., Li, X. (eds.) AI 2018. LNCS (LNAI), vol. 11320, pp. 472–484. Springer, Cham (2018). https://doi.org/10.1007/978-3-030-03991-2_43
8. Durasevic, M., Jakobovic, D.: A survey of dispatching rules for the dynamic unrelated machines environment. Expert Syst. Appl. **113**, 555–569 (2018)
9. Koza, J.R., Poli, R.: Genetic programming. In: Burke, E.K., Kendall, G. (eds.) Search Methodologies, pp. 127–164. Springer, Boston (2005). https://doi.org/10.1007/0-387-28356-0_5
10. Miyashita, K.: Job-shop scheduling with genetic programming. In: Proceedings of the 2nd Annual Conference on Genetic and Evolutionary Computation, pp. 505–512. Morgan Kaufmann Publishers Inc. (2000)
11. Nguyen, S., Zhang, M., Johnston, M., Tan, K.C.: Genetic programming for evolving due-date assignment models in job shop environments. Evol. Comput. **22**(1), 105–138 (2014)
12. Branke, J., Nguyen, S., Pickardt, C.W., Zhang, M.: Automated design of production scheduling heuristics: a review. IEEE Trans. Evol. Comput. **20**(1), 110–124 (2016)
13. Burke, E.K., Hyde, M.R., Kendall, G., Ochoa, G., Ozcan, E., Woodward, J.R.: Exploring hyper-heuristic methodologies with genetic programming. In: Mumford, C.L., Jain, L.C. (eds.) Computational Intelligence. ISRL, vol. 1, pp. 177–201. Springer, Heidelberg (2009). https://doi.org/10.1007/978-3-642-01799-5_6
14. Burke, E.K., Hyde, M.R., Kendall, G., Woodward, J.R.: A genetic programming hyper-heuristic approach for evolving 2-D strip packing heuristics. IEEE Trans. Evol. Comput. **14**(6), 942–958 (2010)
15. Hyde, M.R.: A genetic programming hyper-heuristic approach to automated packing. Ph.D. thesis, University of Nottingham, UK (2010)
16. Bader-El-Den, M.B., Poli, R., Fatima, S.: Evolving timetabling heuristics using a grammar-based genetic programming hyper-heuristic framework. Memetic Comput. **1**(3), 205–219 (2009). https://doi.org/10.1007/s12293-009-0022-y

17. Pillay, N., Banzhaf, W.: A genetic programming approach to the generation of hyper-heuristics for the uncapacitated examination timetabling problem. In: Neves, J., Santos, M.F., Machado, J.M. (eds.) EPIA 2007. LNCS (LNAI), vol. 4874, pp. 223–234. Springer, Heidelberg (2007). https://doi.org/10.1007/978-3-540-77002-2_19

18. Zhang, F., Mei, Y., Zhang, M.: A new representation in genetic programming for evolving dispatching rules for dynamic flexible job shop scheduling. In: Liefooghe, A., Paquete, L. (eds.) EvoCOP 2019. LNCS, vol. 11452, pp. 33–49. Springer, Cham (2019). https://doi.org/10.1007/978-3-030-16711-0_3

19. Zhang, F., Mei, Y., Zhang, M.: A two-stage genetic programming hyper-heuristic approach with feature selection for dynamic flexible job shop scheduling. In: Proceedings of the Genetic and Evolutionary Computation Conference (GECCO), pp. 347–355. IEEE (2019)

20. Durasević, M., Jakobović, D.: Evolving dispatching rules for optimising many-objective criteria in the unrelated machines environment. Genet. Program Evolvable Mach. 19(1), 9–51 (2017). https://doi.org/10.1007/s10710-017-9310-3

21. Hildebrandt, T., Heger, J., Scholz-Reiter, B.: Towards improved dispatching rules for complex shop floor scenarios: a genetic programming approach. In: Proceedings of the 12th Annual Conference on Genetic and Evolutionary Computation, pp. 257–264. ACM (2010)

22. Davis, J.P., Eisenhardt, K.M., Bingham, C.B.: Developing theory through simulation methods. Acad. Manag. Rev. 32(2), 480–499 (2007)

23. Mei, Y., Zhang, M., Nguyen, S.: Feature selection in evolving job shop dispatching rules with genetic programming. In: Proceedings of the 2016 on Genetic and Evolutionary Computation Conference (GECCO), pp. 365–372 (2016)

24. Gilpin, L.H., Bau, D., Yuan, B.Z., Bajwa, A., Specter, M., Kagal, L.: Explaining explanations: an overview of interpretability of machine learning. In: 5th IEEE International Conference on Data Science and Advanced Analytics (DSAA), pp. 80–89 (2018)

Author Index

Printed in the United States
By Bookmasters